Kenneth Maclean

CLIMATIC RESOURCES
AND ECONOMIC ACTIVITY

CLIMATIC RESOURCES AND ECONOMIC ACTIVITY
A SYMPOSIUM

EDITED BY
JAMES A. TAYLOR
Reader in Geography
University College of Wales, Aberystwyth

CONTRIBUTORS

J. A. Taylor
S. J. Harrison
M. B. Alcock, G. Harvey and J. Thomas
J. M. Peacock and J. E. Sheehy
W. H. Hogg
V. B. Proudfoot
J. C. Rodda
A. K. Biswas
I. Burton, D. Billingsley, M. Blacksell, Vida Chapman, Anne V. Kirkby, L. Foster and G. Wall
K. Taylor
P. J. Codling
S. R. Johnson and J. D. McQuigg
W. J. Maunder

DAVID & CHARLES
NEWTON ABBOT   LONDON

0 7153 6496 0

© James A. Taylor and contributors 1974

All rights reserved. No part of this publication may be reproduced, stored in a retrieval system, or transmitted, in any form or by any means, electronic, mechanical, photocopying, recording or otherwise, without the prior permission of David & Charles (Holdings) Limited

Set in 11 on 13 point Times New Roman
and printed in Great Britain by Latimer Trend & Company Ltd, Plymouth
for David & Charles (Holdings) Limited
South Devon House Newton Abbot Devon

List of Contents

		15
	Preface	
1	The Atmosphere as a Resource J. A. TAYLOR	21
2	Problems in the Measurement and Evaluation of the Climatic Resources of Upland Britain S. J. HARRISON	47
3	Measurement, Evaluation and Management of Climatic Resources for Grassland Production in Hill and Lowland Areas M. B. ALCOCK, G. HARVEY and J. THOMAS	65
4	The Measurement and Utilisation of Some Climatic Resources in Agriculture J. M. PEACOCK and J. E. SHEEHY	87
5	The Use of Climatic Information in the Classification of Agricultural and Horticultural Land W. H. HOGG	109
6	The Northern Limit of Agriculture in Western Canada V. B. PROUDFOOT	121
7	Water Resources in the United Kingdom: a Hydrological Appraisal J. C. RODDA	135

LIST OF CONTENTS

8 Application of Mathematical Models to Water Resources Systems Planning   159
   A. K. BISWAS

9 Public Response to a Successful Air Pollution Control Programme   173
   I. BURTON, D. BILLINGSLEY, M. BLACKSELL, VIDA CHAPMAN, ANNE V. KIRKBY, L. FOSTER and G. WALL

10 Some Aspects of the Economics of Air Pollution in the United Kingdom   193
   K. TAYLOR

11 Weather and Road Accidents   205
   P. J. CODLING

12 Some Useful Approaches to the Measurement of Economic Relationships which Include Climatic Variables   223
   S. R. JOHNSON and J. D. MCQUIGG

13 National Econoclimatic Models   237
   W. J. MAUNDER

   Author Index   258

   Subject Index   260

List of Illustrations

Fig	1.1	A generalised vertical section of the atmosphere	22
	1.2	Heat exchanges within the atmosphere, at ground and in the soil/sub-soil	25
	1.3	The oxygen (O_2) and carbon dioxide (CO_2) cycles	31
	1.4	The carbon (C) cycle	31
	1.5	The nitrogen (N) cycle	34
	1.6	The sulphur (S) cycle	35
	1.7	The phosphorus (P) cycle	35
	2.1	Temperature lapse-rates for continental and maritime Europe	52
	2.2	Seasonal variation in lapse-rate—Ben Nevis	54
	2.3	Mean temperature differences between Ynyslas and Brynybeddau, north Cardiganshire	57
	2.4	Effect of altitude on the growth of the moor-rush (*Juncus squarrosus*)	60
	2.5	Relationship of the yield of Sitka spruce (*Picea sitchensis*) to altitude in Taliesin Forest	61
	3.1	Leaf-area index of perennial rye-grass on hill and lowland sites	75
	3.2	Rates of growth of perennial rye-grass on lowland and hill sites	77

Fig	3.3	Effect of shelter and altitude on the growth of perennial rye-grass in spring	80
	3.4	Relationship between rate of growth and leaf-area index of perennial rye-grass	81
	3.5	Effect of shelter and altitude on the growth of white clover in spring	82
	4.1	The comparison of temperatures in a Stevenson screen with temperatures at various levels within a crop of perennial rye-grass	89
	4.2	Temperature profiles within a grass crop	90
	4.3	Comparison of the rate of leaf extension of perennial rye-grass with mean temperature and radiation receipt	91
	4.4	Leaf extension of perennial rye-grass plotted against mean temperatures in the soil and plant canopy	93
	4.5	Profiles of mean soil and air temperatures in a crop of perennial rye-grass (from Peacock (1971))	94
	4.6	Diurnal variations of the mean measured and calculated mid-canopy temperatures, and mean diurnal variations of short-wave radiation	97
	4.7	Crop growth rates of total shoot and root of perennial rye-grass	100
	4.8	Interception of solar radiation and leaf-area index in heated and control plots of perennial rye-grass	101
	4.9	The effect of altering three parameters on predicted crop growth rates	102
	4.10	The effect of species and growth habit in pure grass swards on the crop growth rate and three variables	104
	6.1	Western Canada: mean summer isotherms and extent of settlement	127

LIST OF ILLUSTRATIONS 9

Fig	6.2	Western Canada: mean climatic limits and extent of settlement	129
	7.1	Numbers of rain gauges in the United Kingdom, 1861–1971	140
	7.2	The ground-water network in the United Kingdom	142
	7.3	United Kingdom: 'residual rainfall', estimated at ground level (minus actual evaporation)	146
	7.4	England and Wales: abstractions of water; actual compared with authorised	148
	7.5	England and Wales: total winter rainfall; 100 year return period	150
	7.6	United Kingdom: probable maximum precipitation in 24 hours	152
	7.7	Hydrograph characteristics for the Plynlimon catchments	154
	7.8	England and Wales: forecasted water supply deficits for AD 2001	156
	8.1	General map of the Saint John River basin	164
	9.1	Average smoke concentration near ground level in the United Kingdom (1958–68)	174
	9.2	Average smoke levels at Kew, October–March, 1922–71	175
	9.3	Adoption of smoke control programmes	178
	10.1	Relationship between pollution damage and control costs	194
	13.1	New Zealand butterfat production and weighted 'water deficit' indices, January–April, 1966–70	246
	13.2	New Zealand electricity generation and weighted temperature departures, April–September, 1961–70	248

List of Tables

Table	1.1	Percentage composition of dry air	24
	1.2	The earth's radiation budget	27
	2.1	Percentage of British meteorological stations within sample altitude ranges	48
	4.1	Inputs of the model, and efficiency of conversion of light energy derived from it	99
	5.1	Major climatic controls affecting the environments for plant production	112
	5.2	Macroclimatic restrictions in classification (generalised)	113
	5.3	Definition of climatic groups	116
	7.1	Water resources projects	136
	7.2	Hydrological network classification	137
	7.3	Information on aspects of national hydrological networks	138
	7.4	United Kingdom: river gauging network characteristics	141
	7.5	Estimate of costs of United Kingdom hydrological network	144
	9.1	Variations in the adoption of smoke control areas	177
	9.2	Awareness of air pollution	181

Table	9.3	Perceived seriousness of environmental problems by the general public and by professional sample	183
	9.4	Evaluation of smoke control areas	184
	9.5	Expectation of future episodes of pollution	185
	9.6	Perceived adjustments to pollution	186
	10.1	Source weighting factors for pollution damage costs	198
	11.1	Personal-injury accident data by weather for Great Britain, 1970	209
	11.2	Severity of accidents in relation to weather and road surface conditions for Great Britain, 1970	210
	11.3	Traffic flow changes on the rainiest days during 1969–70 in Great Britain	212
	11.4	Effect of rain (continuous) and wet roads (not raining) on personal-injury accidents at selected places in 1968–9	213
	11.5	Changes in traffic flow and casualties in Great Britain on the six snowiest days during 1969–70	215
	11.6	Personal-injury accidents and reported weather conditions on motorways, 1969–70	217
	11.7	Skidding reported in personal-injury accidents, 1970	218
	12.1	Monthly coefficients for eigenvectors from monthly temperature and precipitation normals	228
	12.2	Average land prices, etc, for counties and climatic variables	230
	12.3	Influences of temperature and precipitation normals on average prices of agricultural land in counties	231
	12.4	Logit estimates of linear functions relating climatically orientated variables to working-day probabilities	234

Table 13.1	Relative significance of data for four geographical counties in New Zealand	241
13.2	Weighted climatic indices for New Zealand	243
13.3	Butterfat processed in New Zealand and associated weighted climatic indices, January–April, 1966–70	245
13.4	Temperature and precipitation departures from the normal in the United States	251
13.5	Retail trade sales in the United States and weighted precipitation and temperature indices	252
13.6	Weighted precipitation and temperature departures and total retail sales for various kinds of business	255

Preface

Academic geography has by tradition afforded some successful communion between scientific and humanitarian studies. Currently, however, increasingly narrow specialisation in individual branches of either physical or human geography has not only seriously impeded this communion but has also created problems of internal communication between physical and human geographers who now frequently find themselves aligned to the research priorities of adjacent fields, eg Quaternary studies, Sociology, etc. The general effects of these trends on academic geography have been divisive. The severe decline in regional geography, the original core of the subject, has become continuous and inevitable.

At the same time, academic geography has recently undergone a major transformation of both method and philosophy. The injection of conventional statistical and mathematical techniques and computer-oriented research has enabled more data to be assembled, processed and interpreted with more precision, and in a manner increasingly acceptable to other sciences. Again, the injection of standard 'systems theory' and 'model building' has enriched the basic philosophy of the subject. More important, these advances have reminded specialist human and physical geographers that they are liable to use the same techniques after all, although the data—and especially the data sources—are taken from different worlds.

Whilst the work of many human geographers has become oriented to the behavioural sciences with a devouring concern to explain human motivation, perception and decision-making processes, the work of many physical geographers, even including geomorphologists, has shifted towards Quaternary studies and the environmental sciences. An expansion of biogeographical, climatological and pedological research has resulted, particularly in its application to environmental or developmental problems in the real world. Thus traditional links between systematic specialisms within geography have been reforged together with the establishment of mutual liaison with the emergent environmental sciences, with which in any event there is much common ground. Environmental materials, processes and events are measured not only *per se* but also in relation to their economic and/or social evaluations and implications. Components of the physical and biological world are assessed in 'resource' terms, both ecological and, ultimately, economic. By the development of studies of the environmental basis of economic activity by biogeographers and applied economists, it is becoming possible to express environmental factors in economic and social terms. This applies more immediately to the environmentally sensitive forms of economic activity, eg agriculture, forestry, construction and power industries, but eventually all industries are affected however indirectly, via transportation, health, etc.

The cycle of deformation and now reformation of centralising, integrative studies within academic geography has coincided with a reawakening, both in public and scientific circles, of the conservation movement in Britain and elsewhere. Questions are now asked about environmental trends and responsibilities as they affect the landscape, the rivers, the soils, the oceans, the atmosphere and, ultimately, human health and planning strategies. This public awareness of environmental trends has given new stimulus and status to biogeographical and associated research and teaching.

This volume of papers, emanating from a symposium convened by the editor under the initial auspices of the Geography Department at the University College of Wales, Aberystwyth on 8–9 March 1972, is therefore timely and representative of current research trends and achievements. The objective is two-fold. First, the atmosphere, its climate and weather, are conceived and measured in resource terms.

Second, the relationships of these climatic resources to specific economic and social usage, which they may subsidise or constrain, are exposed, and so far as possible, quantified so that predictive models may be constructed as aids to the future management and planning especially of weather-sensitive activities. Both objectives are difficult to attain: the first because atmospheric materials and processes have been traditionally measured, *per se* (which is itself difficult enough) rather than in 'applied' terms, thus creating a shortage of data; the second, because so little work in this field has yet been attempted, not least because its value and relevance has only recently been realised. Whilst this volume has the privilege of including contributions from several international authorities in the field of econoclimatology, the major conclusions emerging are the limitations and unsuitability of available climatological, economic and social data, the underdeveloped nature of the subject, and the great potential for future collaborative studies between environmental and social scientists. This applies especially to research into pollution hazards and weather stress as affecting variations in behavioural patterns. Both physical and psychological investigations are necessary to achieve valid interpretations of the relationships involved.

Let it be boldly stated that this volume is *not* a prescription for the revival of any crude type of climatic determinism. On the contrary, it presents an alternative and more sophisticated, interdisciplinary approach to the problem, based primarily on the evaluation of environmental inputs as affecting economic and social decision-making. The rewards are feasible estimates of economic costs and losses and of social inconveniences and preferences. The ultimate rewards are sound prescriptions for environmental management and social planning.

Whether expressed as weather, climate, or atmosphere, the meteorological environment may be scaled on local, national or global terms—as can the parallel economic, social and, ultimately, political responsibilities. The atmosphere, like the soil and the sea, has been taken literally for granted for far too long. Our atmosphere renders this planet unique: it makes life possible. Since man is now technically and potentially capable of contributing to atmospheric changes not only locally and instantaneously but also on a world-wide and permanent basis, the time is right for a re-appraisal of the atmosphere and of its climatic resources for the benefit of all agricultural and industrial activities, and as an aid to social and political adaptations and to environmental management and planning priorities. In the search for ecological responsibilities, the problems are as much educational and cultural as they are technical and scientific in the philosophical atmosphere of an industrialised, urbanised Western society which is traditionally geared to give priority to technological intensification, material gain and economic productivity.

Editor's Acknowledgements
The editor takes pleasure in acknowledging the assistance of many colleagues and students in the preparation of this volume and the original symposium on which it is based. Professor C. Kidson, BA, PhD, Head of the Geography Department at Aberystwyth, allowed access to some departmental services. Professor P. T. Thomas, CBE, PhD, and Mr J. W. Ellis, DRA, Director and Secretary, respectively, of the Welsh Plant Breeding Station, again provided excellent technical and domestic facilities for the symposium. Thanks are due to the University College of Wales and the then Vice-Principal, Professor I. Gowan, BA, MA, for a grant in aid of hospitality. Dr John Cooper, Head of the Developmental Genetics section at the Welsh Plant Breeding Station, kindly and effectively took the chair for some of the symposium papers. My colleagues, Mr D. J. Unwin, BSc, MPhil, of Aberystwyth and Dr A. H. Perry of Swansea, read papers at the meeting on behalf of Dr A. K. Biswas and Dr W. J. Maunder, respectively, who could not be present.

The unqualified success of the symposium was due to the prompt and lively co-operation of speakers and delegates. Attendance totalled eighty-five and included many international representatives. Discussions were energetic and prolonged. Many different academic subjects were represented from a wide range of university institutions. Delegates also attended from several branches of the Scientific Civil Service and the Research Stations. The interdisciplinary drive behind the symposium series was well maintained. For organisational assistance at the symposium itself, a special thank you is due to Miss Janet Davies, BSc, my one-time research assistant and Messrs D. A. Job and M. F. Walker, two of my research students, together with a number of undergraduates.

This was the fifteenth Aberystwyth Symposium in an annual series which began in 1958. The conversion to a two-day meeting from the traditional one-day session, the increasing scale and range of material presented, and the amount of time needed to publish it subsequently in book-form have led to the postponement of the 1973 meeting in anticipation of initiating a biennial series in March 1975, subject to the co-operation of all parties involved.

Finally, I wish to thank the secretarial, cartographic and photographic staff of the Geography Department at Aberystwyth for their continuing and reliable support in preparing material for press.

Aberystwyth
July 1973
J. A. Taylor

CHAPTER 1   J. A. TAYLOR

The Atmosphere as a Resource

Since the end of World War II, research into the structure and processes of the atmosphere has made notable progress with increasing availability of new data for the upper atmosphere, especially the upper troposphere and the stratosphere (Fig 1.1). First, high-flying jet aircraft and radio-sonde balloons and now rockets, satellites and spacecraft have contributed to this technological break-through which has enabled a scientific re-assessment of the atmospheric circulation and the weather and climate it generates. However, parallel to these events, the building of nuclear power stations and the explosion, albeit experimental, of a series of nuclear devices in different parts of the world led to the dumping of radio-active waste in the atmosphere and the oceans to an extent which, although assumed to be controllable and innocuous at first, was ultimately a source of international concern about safety-levels.

The social re-awakening of the conservation movement in Britain and elsewhere in the 1960s was in response to a growing awareness of the pressure of the population explosion and industrial technology on environmental resources. This state of apparent crisis has been usefully studied by Nicholson (1970), for example, but facts have been overused and sometimes exaggerated in both the scientific and popular press due not least to the coincident and accelerated development of powerful mass media. Black (1970) wisely offers valuable corrective in placing the current conservation movement in historical and socio-philosophical perspective. So long as Western society is stringently, nay neurotically, devoted to the pursuit of technological intensification and greater economic productivity, then the ancient biblical image of man having

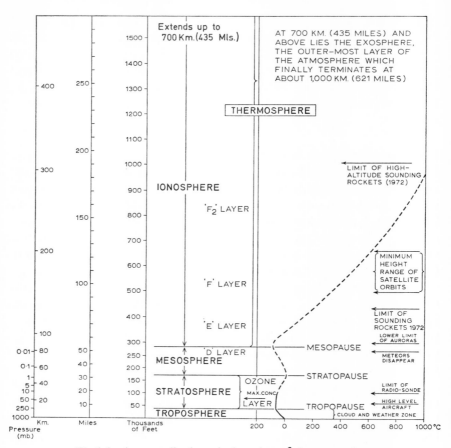

Fig 1.1 A generalised vertical section of the atmosphere

dominion over the earth and having the prerogative to exploit its resources at his convenience, will persist. Inevitably, the social conscience is alerted and the questions of environmental responsibility must be raised. In this context our atmosphere is becoming increasingly perceived in resource terms. This review attempts to describe the atmosphere as a resource in relation to its foundation within the biosphere, its various roles in biogeochemical and nutrient cycling, the impact on it of agricultural and industrial processes, its impingement through climate and weather on economic and social activities, its service and amenity functions and their international implications, and finally, the scope and problems of the proper management of atmospheric resources.

The Atmosphere as an Earth Resource

The earth's oxygenated atmosphere differentiates it fundamentally from other known planetary environments, and for some 2×10^9 years since late Pre-Cambrian times has been the basic life-support system of the biosphere. The geochemical revolution from a reducing-type atmosphere to one containing free oxygen (O_2) created the first differentiated, multi-cellular systems which are ancestral to all subsequent carbon-based life forms as we know them.

The oxygen resources of our modern atmosphere originate from the decomposition of water molecules by light energy in photosynthesis. The primeval oxygen reservoir is believed to have been created by a specialist group of plants which initiated the conversion of the Pre-Cambrian reducing atmosphere. Thereafter organisms had to adapt to the presence of free oxygen to survive and develop.

The total water resources of the earth, estimated at $1.5 \times 10^9 km^3$, are re-cycled via photosynthesis and respiration once every 2 million years or so. After it has entered the atmosphere, oxygen is re-cycled there about once every 2,000 years. Oxygen constitutes just under 21 per cent, by volume, of the atmosphere (Table 1.1). Respired carbon dioxide is added to the small proportion (0·03 per cent by volume) already in the atmosphere; this is in balance with the carbon dioxide found in the oceans and elsewhere in the hydrosphere. Some carbon dioxide is removed from the atmosphere in the precipitation of calcium carbonate in solution. Subsequently formed limestones emerge from the sea and carbon dioxide is returned to the atmosphere as rainfall dissolves the

Table 1.1
PERCENTAGE COMPOSITION OF DRY AIR

	By volume %	By weight %
Nitrogen	78·09	75·50
Oxygen	20·95	23·10
Argon	0·93	1·30
Carbon dioxide	0·03	0·05
Hydrogen, Neon, Helium, Krypton, Xenon, Ozone, Radon	Traces only	

Source: *Handbook of Aviation Meteorology*, MO 630, AP 3340 (1960).

rock surface. Thus, air, land, sea and their several interfaces contribute to the increasingly complex cycling of oxygen in its various combinations with hydrogen and carbon. These geochemical cycles are energised by the basic radiation exchanges (Fig 1.2) in association with nutrient cycling as related to plants, animals and man.

The functioning of the atmosphere as an earth resource is variable but always ultimately vital. It shields, filters, intercepts, reflects, diffuses and absorbs particular wave-lengths of incoming solar radiation; prevents excessive loss of heat energy transmitted by long-wave radiation from the earth; constitutes a reservoir for both short-term and long-term cycling of nitrogen, oxygen, hydrogen, carbon dioxide, water vapour, etc; provides pathways for energy and nutrient cycling and for the translocation of buoyant or wind-borne materials, organic and inorganic; and has become the depository for all kinds of pollutants—gaseous, solid and liquid. Local pollution can be dispersed within the general circulation but it may be transferred for deposition elsewhere, or trapped and intensified below inversions especially when winter anti-

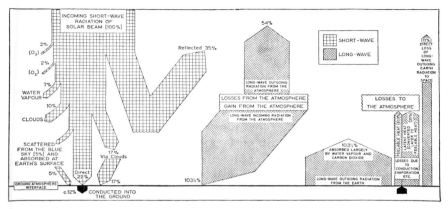

Fig 1.2 Heat exchanges within the atmosphere, at ground and in the soil/sub-soil (adapted from H. H. Lamb, 1972)

cyclones persist. Thus the atmosphere may dilute and remove pollution, but on occasions may act as a source of re-cycled pollutants. Certain notorious materials, eg radio-active waste, DDT, etc, have been comprehensively diffused throughout the entire biosphere as witnessed in the discovery of the presence of DDT in Antarctic penguin populations. The total atmospheric environment is now immediately at the mercy of agricultural and industrial emissions, many of them toxic, and nuclear and other technological experiments, the full implications of which are not fully understood. Hence the need to treat the atmosphere in conservational terms, to balance positive and negative trends, to maintain non-renewable reserves and to control toxicity levels.

The Atmosphere as a Natural Resource

The general structure and vertical layering of the atmosphere are shown diagrammatically in Fig 1.1. Incoming short-wave solar radiation penetrates the atmosphere most effectively when the sun is highest in the sky or, as within the tropics, vertical. Diagonal, longer, and less effective pathways are followed under a low sun. Moreover, the area of ground surface receiving a net radiation beam increases with the obliquity of the pathway. Hence the general latitudinal thermal gradient between the tropics and the poles—a gradient which in turn is modified by land/sea distributions, relief variation and secondary transfers of heat energy within the atmospheric and oceanic circulations.

En route through the atmosphere, about one-third (35 per cent) of the solar beam is reflected back and lost into space mainly from cloud tops (Table 1.2). A second diversion and net loss occurs as some radiation is diffused by air molecules and other substances (eg dust, pollutants and certain biological material such as pollen and spores) held in suspension in the atmosphere. A portion of this diffused radiation is lost to space but more than half eventually reaches the earth. A third process which reduces the initial radiation input from the sun is absorption (i) by the ozone (O_3) layer (2·0 per cent) in the stratosphere at 15–45km (9–28 miles) with a maximum concentration at 22km (14 miles); (ii) by free oxygen (O_2) (2·0 per cent); (iii) by water vapour (7·0 per cent) which is mostly confined to the lower atmosphere below 2·13km (7,000ft); and carbonic acid. It should be noted that the ozone layer intercepts harmful ultra-violet radiation which becomes the main energy source for the circulation above 30km (18·6 miles) and is identified with very high temperatures (Fig 1.1). The energy absorbed in the above three ways is re-radiated within the atmosphere in all directions. The total radiation eventually reaching the earth's surface via clouds is 17·0 per cent of the incoming beam. The proportion of *direct* radiation (insolation) reaching the earth's surface is 22·0 per cent. Some of this is reflected back into the atmosphere and some is in turn reflected back from the base of clouds; this amount, net, is estimated at 8–11 per cent. It emerges that the total short-wave radiation received at the earth's surface from the sun (including 5 per cent scattered from the blue sky) is about 45 per cent of the incoming beam. In addition, however, the amount of radiation via long waves from the atmosphere itself is 90–8 per cent of the incoming beam. Total net radiation received at the sur-

Table 1.2

SUMMARY OF THE EARTH'S OVERALL AVERAGE RADIATION BUDGET: PRESENT-DAY CLIMATES

Incoming			Outgoing		
Absorbed	Transmitted onwards by re-radiation, forward scattering, etc within the atmosphere	Returned Earth radiation	Earth radiation emitted (and lost) to space	Earth radiation from surface to atmosphere	Returned solar radiation
Ozone in stratosphere 2%	Scattered from the blue sky and absorbed at Earth's surface 5–6%	Radiated (long wave) from the atmosphere and absorbed at Earth's surface 90–8%	Direct from Earth's surface 11%	Absorbed largely by water vapour and carbon dioxide 100–7%	Reflected part of solar beam c 35%
Oxygen 2%			Radiated (long wave) from the atmosphere 54%		
Water vapour in lower atmosphere 7%					
Clouds 10%	Absorbed at Earth's surface (from solar beam, ie short wave) after passing through clouds, 17%	Returned to Earth's surface after reflection within atmosphere, 8–11%			
Earth's surface (from direct solar beam), 22%					
Total radiation received at the surface (a) short wave from the sun 44–5% (b) long wave from the atmosphere 98–109% (a) + (b) 142–54%			Total radiation lost from the surface 111–18% Heat lost from surface to atmosphere by conduction, convection, and in evaporation processes, etc (a) directly as feelable heat 6–12% (b) indirectly as latent heat, converted later into feelable heat 20% Total losses from the surface 137–50%		

Units: per cent of the energy available in the solar beam
Source: Lamb, H. H. *Climate: Present, Past and Future. Vol 1 Fundamentals and Climate Now* (1972), p 54

face then is 142–54 per cent of the original input from the sun (Table 1.2 and Lamb, H. H. (1972)).

Terrestrial long-wave radiation emitted by ground, ocean and atmosphere may follow up to four separate pathways. First, it may be reflected back to earth mostly from cloud bases; second, it may be diffused by the same agencies as the incoming radiation but obviously the lower atmosphere components (eg the denser air layers (Fig 1.1), pollutants, etc) have a much more important role to play; third, it may be absorbed in particular by water vapour; and fourth, some may be finally and irrevocably lost to space. Suffice it here to note that the total loss from the surface is 137–50 per cent compared with net receipts of 142–54 per cent of the incoming solar beam (Table 1.2 and Lamb, H. H. (1972)). Thus, in net terms, only about 5·0 per cent of the incoming solar radiation is used at the surface for heating. The so-called atmospheric heat engine is clearly very inefficient but it affords a range and variety of thermal environments which have permitted the evolution and adaptation of the diversity of life forms found in the biosphere.

Oort (1970) quotes a definition of the efficiency of the atmospheric heat engine of less than 1 per cent. On the other hand, the atmosphere fulfils a vital role in the upward transfer of radiant energy from the earth's surface. Without this redistribution there would be a build-up of intense heat at the earth's surface whilst cooling rates in the atmosphere would be of the order of 2° C per day, varying with height. Life would be impossible. Energy is transferred upwards directly from the earth's surface (i) by evaporation from water surfaces, (ii) by conduction and (iii) by long-wave radiation, but a large portion is 'latent' within water vapour, being released to the atmosphere only on condensation. Turbulence, convective cloud formation and mechanical uplift at fronts all contribute to net upward mixing processes which culminate in the tops of the highest cumulo-nimbus clouds at or about the tropopause (Fig 1.1). It is interesting to note that the atmosphere, being warmed from below, is therefore liable to develop vigorous convection. The oceans in contrast are warmed from above which promotes stable conditions and vast lateral circulations driven by the surface winds. At the same time both oceanic and atmospheric circulations co-operate in the advection of heat from the tropics to the relatively heat-deficient middle and high-middle latitudes.

Within the general operation of the atmosphere, the following variables are particularly sensitive in the calibration of spatial and temporal changes in climate and weather:

(a) cloudiness and degree of opacity,
(b) water vapour content,
(c) ozone concentration,
(d) volcanic dust content,
(e) carbon dioxide content.

Cloudiness and degree of opacity modify directly the incoming and outgoing radiation flows by interception, diffusion and absorption. Water vapour is an efficient absorber and radiator of long-wave radiation, and a vehicle for general heat transfer especially in the lower atmosphere. Variation in the ozone concentration has been cited by Allen (1958) as a possible cause of the fluctuation in ultra-violet radiation received between about 1920 and 1940. Other calculations suggest that variations by a factor of 2 or 3 in the amount of radiation intercepted by the stratosphere may be attributed to variations in volcanic dust output since 1880. The effects of the increased CO_2 content of the atmosphere and the possible contribution of industrialisation with consequent rises in temperature have been discussed by Plass (1959). Carbon dioxide is more or less transparent to the visible part of the spectrum but it does absorb invisible infra-red radiation. Now the earth re-radiates much of the infra-red energy, and this process is most intense for wave-lengths between 13 and 17 microns which coincides with the absorption band of the CO_2 spectrum. A high CO_2 concentration will absorb large amounts of infra-red radiation preventing its loss to space and generally raising temperatures.

Thus has been demonstrated the vital role of the atmosphere in the radiation exchange and energy balance of the biosphere. The cumulative availability of net energy resources and their selective deployment through the atmospheric, rock, soil, and oceanic environments constitutes the basic life-line of the biosphere but man is threatening to interfere permanently with the natural equilibrium of the earth's ecosystem.

The Atmosphere as a Biological and Agricultural Resource

Solar energy sustains all living systems. It is 'fixed' by photosynthesis when it is briefly available within the biosphere prior to being radiated back to space. In reality only about 1 per cent of the solar energy received from the sun is used in photosynthesis. Expressed another way, scarcely more than one fine summer day's supply of radiation is actually converted throughout one agricultural growing season (Penman (1968)). Such a vast underuse of the atmosphere's basic resource presents a formidable challenge to agricultural technology since up to 10 or 12 per cent is theoretically available for harnessing.

The atmospheric reservoir of free oxygen (O_2) and the related cyclings of oxygen, water and carbon dioxide render the atmosphere indispensable to the proper functioning of the biosphere. Oxygen also combines with a wide range of other elements in the earth's crust through rock weathering and soil-forming processes. Thus the atmosphere is also essential to the functioning of the lithosphere, the regolith and the soil. In the same context geomorphological evolution has proceeded continuously under the blanket of the atmosphere and its heat and moisture exchanges, and the details of the land surface may owe just as much to the climatic effects of the overlying air as to the structure and character of the underlying rocks.

The oxygen cycle is biologically linked with the carbon dioxide cycle. Both are illustrated in Fig 1.3. Animals take in O_2 and emit CO_2; plants assimilate CO_2 and respire O_2. The role of the atmosphere here is in providing stores for both gases and pathways for their exchange within the biosphere. Even more fundamental is the associated carbon cycle (Fig 1.4). The biological world consists of various combinations of carbon compounds in a continuous state of creation, transformation

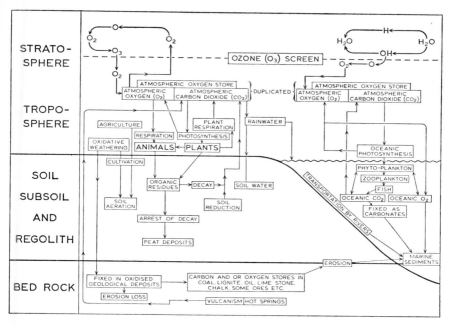

Fig 1.3 The oxygen (O_2) and carbon dioxide (CO_2) cycles

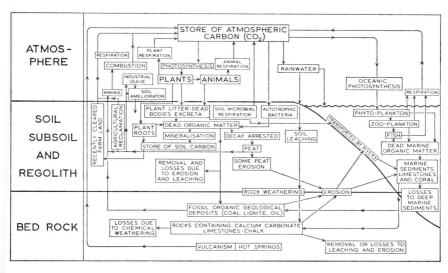

Fig 1.4 The carbon (C) cycle

and decomposition. Again, the reservoir is the atmosphere. Both green plants and phyto-plankton (in the sea) can use the energy in sunlight to transform CO_2 (plus water) into organic molecules of specific structure and ultimately rich variety. Some CO_2 is subsequently returned to the atmosphere by respiration from plant, animal and soil, but about one half of it will be added to the soil in the form of decomposing organic matter. Thus the soil acquires a secondary pool for carbon storage which is very vital to the stability of the soil medium for plant growth. The pathways used by the carbon cycle are several but the crucial factor is the speed with which the cycle is completed. Once every 35 years the CO_2 content of the atmosphere is entirely replaced. Thus man may affect this cycle quickly but in regional rather than global terms, and has done so (a) by extensive deforestation especially in the middle latitudes and (b) by industrial burning especially over the last two centuries in Europe and North America, etc. Over the long term, the burning of wood, coal and oil has, however, not only capitalised on available natural forest resources and on the stored biogeochemical by-products of ancient carbon cycles, but has also returned much carbon to the atmospheric pool. It could mean that much more absorption of outgoing long-wave radiation is taking, and will take, place and temperatures could rise by 2° C by AD 2000 (Plass (1959)). Photosynthesis could also increase and dry matter production by the early twenty-first century could be up by 40 per cent (Jackson and Raw (1966)). The pending deforestation of much of Amazonia, which some Brazilian authorities currently envisage, could contribute regionally to these trends.

The atmosphere is also the basic store for nitrogen on a large scale (78 per cent content by volume) but ironically it is not directly available to all plants. Whilst the reserves of atmospheric nitrogen are virtually inexhaustible, the need to convert them chemically first to nitrites and then nitrates before they can be used in the biosphere gives the means of conversion a supremely important role. Nitrogen must be combined with hydrogen or oxygen before it can be assimilated by plants. The organisms which can 'fix' atmospheric nitrogen are of two types: (1) free-living, eg blue-green algae which obtain the energy required directly from sunlight and (2) those living in symbiotic association with higher plants (about a dozen in number) including the Leguminosae. In the root nodules of the latter, bacteria like *Rhizobium* are instrumental in the nitrogen-fixing process. Obviously, man in planting legumes and applying nitrogenous fertilisers has enhanced a natural cycle for his own agricultural convenience. Moreover, his industrial fixation and production of nitrogen has, since its invention in 1914, achieved a rate four times that operating in contemporary natural fixation (Delwiche (1970)). Currently, plant breeders are developing new species capable of fixing atmospheric nitrogen which could have important implications for agriculture particularly in the food-deficient regions. Small but significant amounts of nitrogen are fixed by cosmic rays, meteor trails, lightning and some marine organisms. Nitrogen may also be added to the soil in falling rain.

The major aspects of nitrogen cycling are indicated in Fig 1.5. The essential point, in summary, is that whilst the atmosphere functions as a major store and pathway, vital agencies, not least man, are required to convert the nitrogen to a usable form. The amount of industrially fixed nitrogen produced is doubling every 6 years. If the nitrogen in fertilisers used in agriculture is added to this then the total fixed by man exceeds by about 10 per cent the total fixed by nature. This excess is moving by run-off into ground-water systems and there is a danger that denitrification will not keep pace with nitrogen-fixing. Denitrification takes place wherever the input of organic materials exceeds the input of O_2 for degradation, eg in the sufficiently anaerobic environments found in the tundra, any swamp or peat bog, or peaty and gleyed soils. It will be essential to maintain long-term balance in the nitrogen exchanges which constitute one of the most complete of the natural cycles of the biosphere.

Much of the nitrogen circulation takes place within the soil, however, which offers an alternative store for many other cycles of both bio-

34 CLIMATIC RESOURCES AND ECONOMIC ACTIVITY

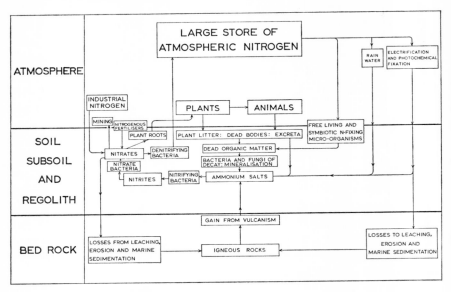

Fig 1.5 The nitrogen (N) cycle

geochemical and nutrient exchange. At the same time the atmosphere provides pathways and temporary stores. The sulphur (Fig 1.6) and phosphorus (Fig 1.7) cycles will serve to illustrate the point. Both elements are derived originally from the soil or weathered layers of the regolith. Both are essential to the growth of living organisms but both are in limited supply in terms of the biological demand.

Under anaerobic conditions, inorganic soil sulphur is converted into H_2S plus sulphides which are then oxidised to produce sulphates (SO_4) in a form available to plants. This process is most effective in well-drained, acid soils and is performed by specialist bacteria, fungi and other organisms, in particular *Thiobacillus thiooxidans*. Alternatively, in heavy textured, waterlogged soils, muds and peats, other specialist bacteria, eg *Desulfovibrio desulfuricans*, reduce the sulphates and release

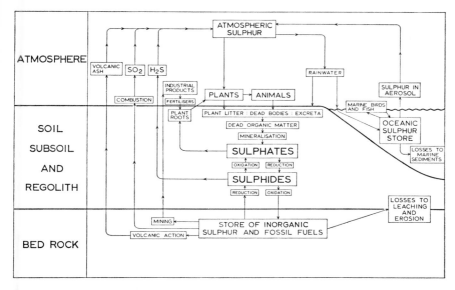

Fig 1.6 The sulphur (S) cycle

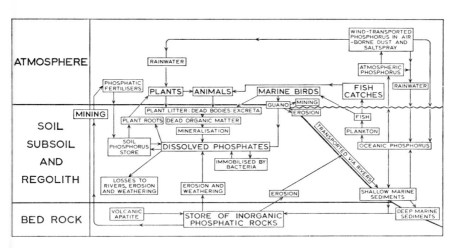

Fig 1.7 The phosphorus (P) cycle

H_2S—which may enter the atmosphere to cycle through rainfall back to the soil. By combustion, industrial SO_2 is also injected into the atmosphere forming temporary pools which may render local pollution very toxic (Jones (1972)).

Inorganic phosphorus is stored in particular geological formations. In the soil both organic and inorganic phosphorus are present in solution to be taken up by plant roots, to enter food chains, eventually to return to the soil by decomposition of organic matter. This cycle is subject to seasonal variations and, with the introduction of phosphatic manures in agriculture, much inorganic phosphorus is eroded from the land by rivers to be deposited in the oceans. Some of this is re-cycled into the atmosphere from the sea surface to reach land again via the wind or salt-spray aerosol. Much more, however, enters the bodies of fish and seabirds, the latter redepositing it in the form of excreta (guano). Some is transferred back to land in fish catches. Odum (1959, 1963) concludes that the net and irrevocable loss of phosphates incorporated in deep ocean sediments is rapidly increasing. Excessive mining of rock phosphate is creating further imbalance. Hutchinson (1954) has estimated that the approximate average annual amount of phosphorus returned to the land in fish catches, etc, is only a tiny fraction (3–6 per cent) of the average annual production of rock phosphate, much of which eventually joins the permanently inaccessible store in the sea. The prospect of phosphorus deficiency in many parts of the biosphere looms large.

These cycles illustrate the varying roles of air, sea, rock and soil as stores, large or small, temporary or permanent, for the basic elements and nutrients, and the varying complexity of the pathways they provide. The role of man is equally complicated. He may accelerate or decelerate, intercept or ignore, dilute or strengthen, halt or reverse, imitate or accept, use or abuse, exploit or conserve selected interchanges. All biogeochemical balances are ultimately affected by the direct and indirect consequences of these several alternates. The atmosphere emerges as a major store and versatile pathway. It functions as a positive resource for flora, fauna and agriculture by providing adequate energy inputs in the flow of light, heat and water. However, where these inputs are individually or collectively inadequate or occur to unwanted excess, the atmosphere via climate thresholds and weather hazards imposes constraints on specifically sensitive biological or agricultural activities. It depends ultimately on the requirements of individual organisms or systems which, by adaptation, will evaluate the atmospheric resources available. When the relationship between organism and climatic environment is correct, productivity will vary according to many factors but including the fluctuation in atmospheric resources for individual seasons.

The Atmosphere as an Economic Resource

All economic activity is dependent in general on supplies of natural energy ultimately derived from the sun. This includes not only agriculture, forestry, etc, which are located outside in the soil/air contact zone of the biosphere but also 'indoor' manufacturing and all forms of industry especially those which process natural materials (eg food processing, pulp and paper-making, textiles and metal working, etc) or synthetic products (eg artificial fibres, plastics, etc) which are still ultimately dependent on the primary and finite stores of organic and inorganic parent materials as available in the biosphere. The measurement of atmospheric resources for economic purposes, however, must reconcile the standard requirements of instrumentation with the particular requirements of the economic activity concerned. For example, Meteorological Office procedure is based on the standardisation of meteorological instruments, their recording range and exposure. The conventional instrument, however, from the economic or applied viewpoint, may be measuring the wrong parameter at the wrong point. Its range and thresholds may not compare with the spectrum of the biological or economic activity or process.

Rodda (1970) has shown for instance that rainfall data relating to conventional gauges 30·5cm (12in) above ground may underestimate rainfall at ground by as much as 10 per cent. Rainfall at ground is difficult to measure but this is nearer to the actual water input being received at ground. When an error of that magnitude is extrapolated to all hydrological cycles, serious problems arise when the data are applied to water resource predictions. Again, conventional scales of temperature should be converted to biological and agricultural terms, assuming that the instrumental exposures are within the crop or animal environment concerned and not in an obtruding Stevenson screen, for example. Calculations of aggregated data above some established biometeorological threshold are more useful than simple arithmetic means or ranges but it is essential to discover the extent to which plant and crop responses are not linear. The point is clearly shown with reference to lapse-rates of temperature with altitude (Harrison herein, Chapter 2).

Technology enables the correction of excessive or inadequate climatic inputs, eg irrigation in arid lands, shelter-belts on windy sites (Caborn (1957, 1964)), etc. The plant and animal breeders may develop new varieties or types to extend a climatic frontier or tolerate a climatic constraint, eg the vernalisation of wheat seed to accommodate the very short growing season in the northern prairies of Canada (Proudfoot herein, Chapter 6); the breeding of new coffee varieties in Brazil which will be more resistant to coffee leaf-rust disease and, optimistically, frost (Taylor (1974)); the recent introduction of short-stem rice (I.R.8) which promises to revolutionise production in many food-deficient lands but which is itself more vulnerable to disease than was originally anticipated. Many biological hazards are related in occurrence and spread to specific meteorological conditions, eg potato blight, liver fluke, etc—see Taylor, ed (1962), Hogg (1967), Ollerenshaw (1966, 1967, 1971). The wind in particular has been identified as a vector in the spread of many diseases, including 'foot and mouth' (Tinline (1969); Smith and Hugh-Jones (1969)). Climatic resources and weather and related hazards affect the location and productivity of agricultural enterprises to varying degrees. Climatic constraints for heavy crops, etc, imply low yields and, usually, low profitability. Abundant availability of climatic resources implies high agricultural yields and, usually, high profitability. Moreover, protection against hazards costs money.

Many industrial activities are also variably sensitive to weather and climate, indirectly or directly. Sewell *et al* (1968), Maunder (1970), Taylor, ed (1970, 1972) and McQuigg (1972) have revealed the full scope of the relationship between atmospheric resources and their economic and social evaluation and management. Weather sensitivity may be expressed in the quantity or quality of raw materials, their processing, and in transportation and labour costs. The extent to which an economic activity is exposed to weather stress is approximately proportional to the additional costs accruing. Rarely, however, is the equation available because of lack of precise data. In many instances, entrepreneurs are unaware of weather costs which are often assumed to be negligible, unavoidable and therefore unimportant.

In reality the adjustment of decision-making and planning strategies in the light of weather information (including forecasts) can achieve substantial economies in the operation of selected agricultural, industrial, social enterprises—see Curry (1962), Duckham (1967), Stansfield (1970), Stringer (1970), Taylor, ed (1970, 1972), Thomas, W. L. (1972). Buchanan (1972) has revealed the extent to which particular British industries depend on meteorological advice—including the power, construction, transportation and food and beverage industries. Unfortunately, many weather costs remain latent, unperceived and unaccounted; they are carried and tolerated in aggregate turnovers. The national, and even the local, costs must be formidable if the problems in extracting accurate data and calculating losses could be overcome. Both Johnson and McQuigg (herein Chapter 12) and Maunder (herein Chapter 13) attempt a statistical assessment of econoclimatic variables as mirrored in local (county) or national data for eg land prices, road construction projects, electricity production, dairying output and, finally, weekly retail trade sales. The inadequacy and crudeness of aggregate data is apparent and presents problems of interpretation. Far too many intermediate, unmeasured and often imponderable factors intervene between the weather inputs and the ultimate economic, social or managerial consequences (Taylor (1972)).

The Atmosphere as a Social Resource
Our perception of the atmosphere is changing. Once taken for granted as an allegedly infinite envelope of gases around our planet, the atmosphere has now been intensively explored as never before and can be studied from below, within and above. Its evaluation as a resource of all kinds, not least social and also, inevitably, political, is achieving new dimensions and new meanings. As a source of light, heat, sunshine and plain 'fresh air', the atmosphere has massive amenity and leisure value. Its major rivals within the framework of the earth's biogeochemical cyclings are the hydrosphere, the lithosphere and the regolith/soils mantle. However, as an energy source, store and pathway it is major and unique.

Moreover, the atmosphere is the primary conditioner of man's created 'indoor' environments as applied to the house, office, factory or city, and the costs of establishing and maintaining these environments. Architectural adjustments in terms of plans and buildings materials should accommodate the incident climate by the enhancement of basic resource supplies, eg adjustment of windows to light, heat and humidity variations, and the avoidance of the over-exposure or extra exposure created by badly planned buildings, eg predominant wind stress and wind funnelling. Such sophisticated anticipation of the climate of built-up areas requires much more direct instrumentation and data (Page (1972)).

During the Industrial Revolution and with the increasing urbanisation of the present century, the atmosphere has come to be treated as an open sewer into which effluents and pollutants of all sorts and sizes have been indiscriminately dumped. Landsberg (1941) reported that the concentration of particles having a radius of less than 0·2 microns (the so-called Aitken nuclei) was 16 times greater on average over cities than over the countryside, and 160 times greater than over the sea.

This indicates the severity of the pollution gradient as it was 35 years ago. Whilst it is essential to monitor and, if possible, prevent the occurrence of physical pollution by the introduction of smokeless fuels, smokeless zones, and by forcing industry to cater properly for the disposal of its own effluent, it is equally essential to educate the general public, as well as the professional scientists, industrialists and administrators concerned, to the true scale of the problem. Burton *et al* (herein Chapter 9) are studying the perception of pollution in a selection of cities on both sides of the Atlantic. Adequate liaison between physical and perception studies of pollution will be necessary to achieve a full understanding of occurrence and impact. Dramatic events like the December 1952 smog which was responsible for the death of 4,000 people in the city of London are a great stimulus to advancing perception and official reaction. Motorway pile-ups due to fog banks make a similar point which should not be ignored. The health hazards, eg, bronchitis, in polluted areas have been usefully mapped for the UK, by Howe (1963).

The amenity role of the atmosphere via weather has been studied by Paul (1972) and Perry (1972). The direct relationship between ascending air currents and gliding has been studied by Scorer (1958); it is a good example of direct amenity use of a special component of the atmosphere at specifically suitable sites. The growing networks of air transport have led to the designation of air corridors especially in the intensely used areas, eg the North Atlantic, USA and Europe. The political and military significance of 'national' air space has sharpened with the increasing sophistication of air travel.

The atmosphere is global and international. It belongs to the biosphere and is accessible to plants, animals and man. It constitutes a resource to the earth, to plants and animals, and for the economic and social activities of man. Man emerges as a powerful agent of change in all these resource relationships. He may intentionally manipulate local weather by cloud-seeding; he may inadvertently create clouds in the stratosphere from the vapour trails of his jet aircraft; he can make pollution or disperse it; he may dream of harnessing solar energy one day more effectively than at present. But he will never match the scale and turnover of solar energy. One day's input of the latter is equivalent to the energy released by 100,000 hydrogen bombs (of the type exploded in 1954) and 10,000 hurricanes. Alternatively, this daily input is equivalent to about 200 times man's annual energy consumption (Chandler (1970)).

None the less, the responsibility for managing atmospheric resources is entirely our own. Some system of international surveillance of those processes currently affecting resource cycling is needed lest our atmospheric heritage becomes irrevocably depleted by uncontrolled dumping of toxic materials. The basic problem, however, is as much educational, social and political as it is technical, biological and scientific.

Acknowledgements
Grateful acknowledgements are due to W. H. Hogg, for constructive comments on the text, and Dr Ellis Griffiths and David B. James (Department of Agricultural Botany, University College of Wales, Aberystwyth) for assistance with Figs 1.3–1.7 inclusive which have been adapted from several sources, including Odum (1959, 1963), Jackson and Raw (1966) and Watts (1971).

References
ALLEN, C. W. 'Solar Radiation', *QJ Roy Met Soc*, **84** (1958), 307–18.
BLACK, J. *The Dominion of Man: The Search for Ecological Responsibility* (Edinburgh, 1970).
BUCHANAN, R. A. 'Weather Forecasting for Industry', in Taylor, J. A. (ed), *Weather Forecasting for Agriculture and Industry* (Newton Abbot, 1972), 115–25.
CABORN, J. M. *Shelterbelts and Microclimate*. Forestry Commission Bulletin No 29 (Edinburgh, 1957).
——. *Shelterbelts and Windbreaks* (1964).
CHANDLER, T. C. 'The Management of Climatic Resources', inaugural lecture, University College London (1970).
CURRY, L. 'The Climate Resources of Intensive Grassland Farming: The Waikato, New Zealand', *Geog Rev*, **52** (1962), 174–94.

DELWICHE, C. C. 'The Nitrogen Cycle', *Sc Amer*, **223**, No 3 (1970), 136–46.
DUCKHAM, A. N. 'Weather and Farm Management Decisions', in Taylor, J. A. (ed), *Weather and Agriculture* (Oxford, 1967), 69–80.
HOGG, W. H. 'The Use of Upper Air Data in Relation to Plant Disease', in Taylor, J. A. (ed), *Weather and Agriculture* (Oxford, 1967), 115–27.
HOWE, G. M. *National Atlas of Disease Mortality in the United Kingdom* (1963).
HUTCHINSON, G. E. 'The Biogeochemistry of the Terrestrial Atmosphere', in Kniper, G. P. (ed), *The Earth as a Planet* (Chicago, 1954), 371–433.
JACKSON, R. M. and RAW, F. 'Life in the Soil', *Studies in Biology*, No 2, Institute of Biology (1966).
JONES, G. E. 'An Investigation into the Possible Causes of Poor Growth of Sitka Spruce (*Picea sitchensis*)', in Taylor, J. A. (ed), *Research Papers in Forest Meteorology: An Aberystwyth Symposium* (1972), 147–56.
LAMB, H. H. *Climate: Past, Present and Future. Vol I Fundamentals and Climate Now* (1972).
LANDSBERG, H. E. *Physical Climatology* (Pennsylvania, 1941 and later editions).
MAUNDER, W. J. *The Value of Weather* (1970).
MCQUIGG, J. D. 'Simulation Model Studies of the Impact of Weather on Road Construction and the Movement of Heavy Equipment in Agricultural Operations', in Taylor, J. A. (ed), *Weather Forecasting for Agriculture and Industry* (Newton Abbot, 1972), 147–54.
NICHOLSON, E. M. *The Environmental Revolution* (1970).
ODUM, E. P. *Fundamentals of Ecology* (Philadelphia, 1959; 2nd edn).
——. *Ecology* (New York, 1963).
OLLERENSHAW, C. B. 'The Approach to Forecasting: The Incidence of Fascioliasis over England and Wales, 1958–62', *Agr Meteorol*, **3** (1966), 35–53.
——. 'Climatic Factors and Liver Fluke Disease', in Taylor, J. A. (ed), *Weather and Agriculture* (Oxford, 1967), 129–35.
——. 'The Influence of Climate on the Life Cycle of *Fasciola hepatica* in Britain with Some Observations on the Relationship between Climate and the Incidence of Fascioliasis in the Netherlands', *Symposium volume, Centraal Diergeneeskundig Instituut* (Lelystad, 1971), 41–63.
OORT, A. H. 'The Energy Cycle of the Earth', *Sc Amer*, **223**, No 3 (1970), 54–63.
PAUL, A. H. 'Weather and the Daily Use of Outdoor Recreation Areas in

Canada', in Taylor, J. A. (ed), *Weather Forecasting for Agriculture and Industry* (Newton Abbot, 1972), 132-46.

PAGE, J. K. 'The Problem of Forecasting the Properties of the Built Environment from the Climatological Properties of the Green-field Site', in Taylor, J. A. (ed), *Weather Forecasting for Agriculture and Industry* (Newton Abbot, 1972), 195-208.

PENMAN, H. L. 'The Earth's Potential', *Sc J*, **4,** No 5 (1968), 43-7.

PERRY, A. H. 'The Weather Forecaster and the Tourist—the Example of the Scottish Skiing Industry', in Taylor, J. A. (ed), *Weather Forecasting for Agriculture and Industry* (Newton Abbot, 1972), 126-31.

PLASS, G. N. 'Carbon Dioxide and Climate', *Sc Amer*, **201** (1959), 41-7

RODDA, J. C. 'On the Questions of Rainfall Measurement and Representativeness', *Proc Symp World Water Balance*, IASH/UNESCO, Reading, IASH Pub No 92 (1970), 173-86.

SCORER, R. S. *Natural Aerodynamics* (Oxford, 1958).

SEWELL, W. R. D. *Human Dimensions of the Atmosphere*, National Science Foundation (Washington DC, 1968).

SMITH, L. P. and HUGH-JONES, M. E. 'The Weather Factor in Foot and Mouth Disease Epidemics', *Nature*, **223** (1969), 712-15.

STANSFIELD, J. M. 'The Effect of the Weather on Farm Organisation and Farm Management', in Taylor, J. A. (ed), *Weather Economics* (Oxford, 1970), 11-16.

STRINGER, E. T. 'The Use of Cost-benefit Studies in the Interpretation of Probability Forecasts for Agriculture and Industry: An Operational Example', in Taylor, J. A. (ed), *Weather Economics* (Oxford, 1970), 83-91.

TAYLOR, J. A. (ed). *Climatic Factors and the Incidence and Spread of Pests and Diseases in Plants and Animals*, Memorandum No 5, Geography Department, UCW (Aberystwyth, 1962).

——. *Weather Economics* (Oxford, 1970).

——. *Weather Forecasting for Agriculture and Industry* (Newton Abbot, 1972).

——. 'The Revaluation of Weather Forecasts', in Taylor, J. A. (ed), *Weather Forecasting for Agriculture and Industry* (Newton Abbot, 1972), 1-18.

——. 'Some Current Problems in Brazilian Coffee Production', *Geografisch Tijdschrift Nieuwe Reeks*, No 1 (1974).

THOMAS, W. L. 'The Value and Relevance of Weather Study and Weather Forecasting in the Profitable Production of Early Potatoes', in Taylor, J. A. (ed), *Weather Forecasting for Agriculture and Industry* (Newton Abbot, 1972), 86-98.

TINLINE, R. 'Meteorological Aspects of the Spread of Foot and Mouth Disease: Evidence from the 1967/8 Epizootic in the English Midlands', *Biomet*, Pt 2, IV (1969), 102.
WATTS, D. *Principles of Biogeography* (1971).

CHAPTER 2   S. J. HARRISON

Problems in the Measurement and Evaluation of the Climatic Resources of Upland Britain

Introduction

In a geographical context, climate possesses three basic co-ordinates of spatial variation, these being (i) latitude, (ii) distance from the sea and (iii) elevation. Depending on the spatial scale, these co-ordinates assume varying degrees of prominence and complexity according to their degree of modification by local climatic factors, in particular, topography. In Britain, the problem of interpreting changes in the climatic environment due to increases in altitude is complex for four major reasons.

Firstly, an increase in altitude is usually accompanied by a greater diversification of relief. Estimates of typical climatic conditions as representative of particular elevations are confounded by variations in slope, aspect and relative relief, all producing small-scale climatic differentiations.

Secondly, climatic data, for a sufficient range of elevations, are not as yet available in Britain. Table 2.1 shows the proportion of stations occurring within sample altitude ranges. The proportion above 300m is only 4·1 per cent for Britain as a whole. This, and the fact that about one-fifth of the land surface of Britain lies above 300m, clearly demonstrates the inadequacy of the meteorological network upon which estimates of the effects of altitude on climate must be made.

With specific reference to Wales, 28 per cent of the land surface is higher than 300m above sea-level but this zone is served by 11·3 per cent of the total number of meteorological stations in the Principality.

Table 2.1

PERCENTAGE OF BRITISH METEOROLOGICAL STATIONS WITHIN SAMPLE ALTITUDE RANGES

Altitude range in metres above sea-level	England	Wales	Scotland	Britain
0 to 150	89·5	69·8	75·2	83·1
150 to 300	8·3	18·9	19·4	12·8
300 to 450	1·3	11·3	3·6	3·0
450 to 600	0·6	0·0	0·0	0·4
Above 600	0·3	0·0	1·8	0·7

Source: *MO Monthly Weather Report* (January 1969).

Isolated research projects, however, have done much to improve the state of knowledge of the local effects of altitude on climate in Britain. The Scottish Meteorological Society's records for the Ben Nevis Observatory (1,343m OD) for the period 1884–1903 (Buchan et al (1905–10)) and Manley's pioneer work in association with the Dun Fell Observatory in the northern Pennines (897m OD) (Manley (1942)), are noteworthy.

Thirdly, there has always been a tendency to over-exaggerate the division between 'lowland' and 'upland' climates without successfully determining a boundary between the two. Miller (1931) had the problem of fitting his 'Mountain Climate' into a rigid global classification, and succeeded only in assigning arbitrary and subjective boundaries. Although not on the same scale as Miller's classification, references to 'mountain climate' (Pearsall (1950)) and 'upland climate' (Oliver (1960)) have frequently been made in Britain, usually on the basis of subjective criteria. Climate is continuous in time and space: variations in climate due to increases in altitude are continuous, and the insertion of arbitrary divisions within these gradients inhibits a full understanding of the climatic effects of altitude.

Fourthly, the other co-ordinates of climate tend to confuse the operation of the altitude factor. The rate of climatic deterioration with altitude differs, for example, as between west- and east-facing slopes and between the west and east coasts of Britain. The bioclimatic maps of Scotland produced by Birse (1971) clearly indicate a more rapid altitudinal change in climate in the west as compared with the east (*ceteris paribus*). Both Manley (1945) and Taylor, ed (1960) have contrasted the steeper temperature lapse-rates in maritime uplands with the gentler ones of continental mountain areas.

Growth Gradients

A relationship exists between the rate of plant growth and the amount of energy available in various forms in the climatic environment. Changes in the biological resource value of climate along its three spatial co-ordinates may be referred to as 'growth gradients'. A number of climatic variables exhibit gradients of change with altitude and, individually and collectively, are of direct relevance to plant growth. These include temperature, precipitation, snow-cover, wind speed, cloud cover and insolation. Investigation into these gradients, based on the Ben Nevis data, has shown that decreased temperatures and sunshine hours, and increased precipitation and wind speed generally accompany increase in surface elevation.

These climatic gradients have occasionally been expressed in hypothetical terms related more directly to plant growth. Accumulated temperature above a standard growth-threshold (eg $6°$ C) (Shellard (1959)), for example, is considered a more sensitive measure of heat availability than temperature alone. The amount of information available, however, is insufficient to allow accurate assessment on such terms.

Since adequate supplies of both moisture and heat are prerequisites for plant growth, precipitation increases with altitude may be regarded as a positive, rather than a negative component of a general decrease in the potential for plant growth with increasing altitude in Britain. Temperature, normally decreasing with altitude, must be regarded as negative along such a gradient, and it is this factor that has attracted most attention, particularly through the work of Manley (1945) who interpreted temperature records through an assessment of a hypothetical growing season. The extreme climatic environment of highest elevations in Britain is most clearly expressed as a thermal differentiation, but the absence of an adequate meteorological recording network for these zones has made necessary the use of estimating techniques, the efficacy of which must be questioned.

Lapse-rates
A theoretical approximation to the decrease in temperature with elevation has been termed the temperature lapse-rate, and has been widely applied to variations into the free atmosphere and over rising ground surfaces. In Britain, a standard lapse-rate in mean temperature of 6° C for every 1,000m increase in altitude has been accepted as a general estimate of the decreases recorded in the Stevenson screen. This has been widely used (a) as a means of reducing mean temperatures to a common base-level to facilitate the isolation of other factors affecting the distribution of temperature, (b) for inserting temperature records at intermediate altitudes where such are not available and (c) for the extrapolation of the same to higher altitudes.

The employment of the standard lapse-rate for such purposes is open to criticism, not least that the degree of approximation involved produces results which are below the acceptable level of accuracy. If the standard lapse-rate is to be utilised in this way, ie as representative of an altitudinal temperature decrease along a growth gradient, then the following four fundamental problems must first be overcome:

1 Spatial Variation in the Effect of Altitude upon Temperature
Manley (1945) and Taylor, ed (1960) compared simple decreases in the length and intensity of a thermally defined growing season, which served to demonstrate the severity of the impact of altitude-induced temperature changes in Britain in contrast to both continental Europe and North America. A calculated reduction of 10 days in the length of the growing

season for every 84m increase in altitude for the southern Pennines was contrasted with 10 days for every 155m in southern Vermont (Manley (1945)).

Using lapse-rates selected from several published for Europe, a simple comparison can be made between the standard estimate for Britain of 6·0° C per 1,000m and Hess's (1968) values for the Carpathians which range between 4·3 and 5·5° C per 1,000m—which emphasises the scale of the contrasts originally indicated by Manley (Fig 2.1). More important, within Britain regional differences arise, a lapse-rate in mean annual temperature of 6·9° C per 1,000m having been derived by Manley (1943) for the northern Pennines, 6·7° C by Smith (1950) for west central Wales, and 7·3° C by Oliver (1960) for South Wales. However, spatial variations within the British Isles have not been adequately investigated and the isolated examples quoted here are totally insufficient to allow a conclusive statement to be made, not least because of a number of anomalous siting factors at some of the meteorological sites adopted. Manley (1945) has contrasted maritime and continental locations demonstrating the steeper lapse-rates in the former, and states: 'It is evident that rapid change with altitude in the length of the growing season ultimately depends upon the frequency with which maritime polar air masses invade the given region.' To some degree, this simple contrast operates within Britain between west and east coasts (Birse (1971)). Postulated spatial variability in the regional and local effects of altitude upon temperature in Britain may, therefore, severely restrict the applicability of a standard lapse-rate based on national averages and mean conditions.

2 Temporal Variation in the Rate of Change in Temperature with Altitude

Temporal variability across a spectrum of time-scales, is a major problem in the calculation and applicability of mean annual lapse-rate of temperature in Britain, where daily and seasonal conditions are changeable. Manley (1943) and Oliver (1960) have clearly demonstrated that there exists some degree of seasonal variation in the

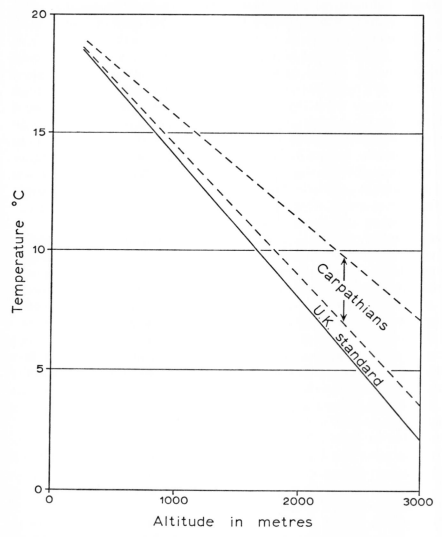

Fig 2.1 Temperature lapse-rates for continental and maritime Europe

effects of altitude on temperature, and it has been suggested that lapse-rates are steepest in *spring*. Further analysis of a synopsis of the Ben Nevis observations (Pearsall (1950)) also confirms the point (Fig 2.2).

From March 1969 to March 1971 screen temperatures were recorded by the author at two stations in north Cardiganshire, Wales. These were (a) at Ynyslas, 3m above sea-level and virtually on the Cardigan Bay coast, and (b) at Brynybeddau, at 450m, on the western slopes of Plynlimon Fawr some 20km inland. The four-weekly differences in mean screen temperature, presented in Fig 2.3, were at a maximum in *late autumn* and *winter* for the period of the investigation. This confirms the findings of Smith (1950), who analysed available mean temperatures in the same area, and concluded that the fall in temperature with height was *least* in *spring*, especially in *March* and *May*.

The cause of this seasonal variation in lapse-rates has never been satisfactorily explained. Manley (1942) suggested that the varying frequency of polar air masses was the source. Reservation must be placed upon this simple correlation as screen temperatures are affected by both atmospheric and ground surface conditions. The experiments in north Cardiganshire, in 1969 and 1970, revealed that there was some coincidence between the occurrence of cold polar airstreams and larger temperature differences between the two stations. This was by no means exclusive, however, very marked differences arising (a) when snow was lying at the upper station and (b) when there was great disparity in the cloud cover at the two.

Smith (1952), has also emphasised the divergence between lapse-rates as measured in mean daily *maximum* and *minimum* temperatures, those for the former frequently being steeper than those for the latter. To apply the standard lapse-rate, therefore, is to underestimate the scale of variations occurring over short and long time periods.

3 The Derivation of Lapse-rates

Nearly all British lapse-rates have been constructed mostly on the basis of two terminal recording points. Manley (1943), for example, managed

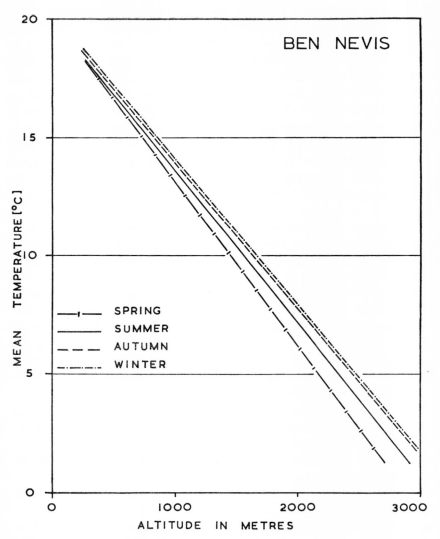

Fig 2.2 Seasonal variation in lapse-rate—Ben Nevis

to produce lapse-rates between paired stations of Moorhouse and Newton Rigg to the west, and Moorhouse and Durham to the east, of the Pennines.

Two basic errors are involved here. The first is in the local siting characteristics of the climate stations used, as diversity of surface relief will generate deviations in local climate which detract from the 'representativeness' of the station data as applied to its specific altitude. The derivation of lapse-rates from only *two* reference points may be tantamount to comparing temperatures which originate from local rather than altitudinal factors. Oliver (1960), for example, admits that a relationship between altitude and temperature in South Wales may have been conditioned by local siting factors at the stations he used.

The second is the error in assuming that the rates of change in temperature between these two points are uniform. Although the linear form is generally accepted, the possibility of, for example, exponential or sigmoidal forms must not be dismissed.

Reference to just two climate stations cannot give a true indication of the nature of meteorological changes between them but can merely express a temperature difference in terms of their altitudinal separation which may be due to a combination of altitude and local topography. Thus, although a limited number of regionally determined lapse-rates have been estimated, many have dubious value because of site errors and oversimplified construction.

4 *The Level at which Observation of Temperature is Taken*

Lapse-rates have been calculated using conventional temperature records from the Stevenson screen, at 1·2m above the ground surface. If a meaningful measure of growth gradients is to be obtained, then temperatures relating to an air layer which many plants, particularly grasses, do not normally penetrate, are not directly appropriate. Of more relevance are temperature observations in the air layer immediately above, or in the soil immediately below, the ground surface.

Within this restricted zone, the degree of meso- and microclimatic modification is at a maximum where properties of surface and sub-surface, such as vegetation and soil type, have the greatest effect, and the role of altitude is, therefore, more difficult to gauge.

One of the most important factors controlling climatic variation within the zone of the microclimate (Geiger (1965)) is the nature of the soil. Increases in altitude in Britain often result in greater depths of semi-decomposed organic matter in the soil surface horizons, culminating in peaty podzols and upland peats. Consequently, distinct altitudinal differences occur in the thermal properties of soils, increases in the extent and thickness of surface peaty layers normally resulting in a lower thermal conductivity. Taylor (1967) has indicated the scale of contrasts in the thermal capacities of organic peaty soils and dominantly inorganic sandy soils on the plain of south-west Lancashire where the microclimatic modification of the onset, intensity and length of the growing season has repercussions on farming decisions affecting the length or shortness of rotations and the planting of sensitive early crops. Lapse-rates in temperature calculated from the Stevenson screen data will, therefore, be confounded by more significant variations occurring nearer the surface, arising mainly from differences at that surface and in the soil.

In the north Cardiganshire experiments conducted by the author, records from the screen were supplemented by observations of temperatures at 20cm above the ground, and at 5 and 10cm below the soil surface. The soil at the Ynyslas station, formed from dune sands, contrasted markedly with the peaty upland soil at the Brynybeddau station. The respective percentages of organic matter by weight in the 10cm immediately below the surface were 3·7 and 84·1. Screen temperature differences between the two stations often bore little relation to differences at other measurement levels. Particularly divergent was the seasonal variations in soil temperature differences which exhibited a very marked and regular sequence of summer maximum and winter minimum. The pattern is clearly affected by contrasts in the thermal properties of the

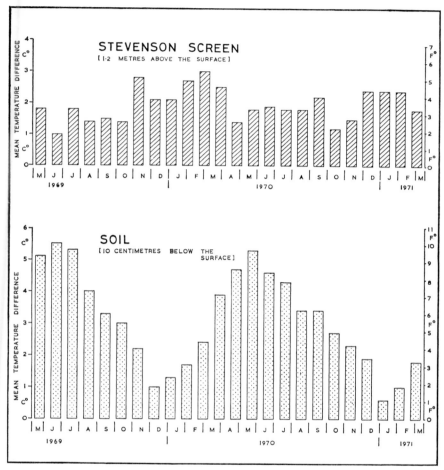

Fig 2.3 Mean temperature differences between Ynyslas and Brynybeddau, north Cardiganshire

two soils at different seasons (Fig 2.3), the major factor being the deeper heat penetration in the Ynyslas sands compared with the upland podzol at Brynybeddau.

More than the basic screen record is required if a true assessment is to be made of an altitudinal growth gradient as defined in climatic terms, a detailed analysis of microclimate being desirable. Because of the intensity of small-scale climatic variation this solution must be regarded as an ideal but certainly an improved soil temperature recording network, for example, would alone give more validity to the use of lapse-rates in assessing growth gradients.

The Validity of a Standard Lapse-rate
The four problems outlined above must be regarded as imposing serious limitations on the use of the simple temperature lapse-rate in its conventional form. It is doubtful whether the standard lapse-rate of 6° C in mean temperature per 1,000m is a valid approximation or whether it has much applicability. Considering the inherent variability of British weather and climate, it may be wiser, until more detailed analyses are available, to adopt local approximations on small spatial and temporal scales for the estimation of altitudinal growth gradients. Emphasis need not be placed entirely upon temperature, for many of the problems outlined, especially the scalar problems, can, in principle, be extended to other climatic variables where there has been an equally inadequate expression, often by crude extrapolation, of the continuum of change with altitude.

The Relevance of Climatically-Defined Growth Gradients
If climatic data existed which permitted solutions to the basic problems outlined above, and which gave an accurate climatic expression of growth gradients, the application of such assessments would still be open to question. Climate forms only a part of the plant environment and, in this respect, to regard it in isolation would not be justifiable.

The application of the climatically-defined growth gradient is restricted unless its exact environmental role be determined. Examples can be drawn from both non-cultivated and cultivated plant growth.

Pearsall (1950) investigated the relationship between the growth of the moor-rush (*Juncus squarrosus*) and altitude, and determined that seed and flower production, and vegetative growth were all diminished at higher altitudes, Fig 2.4. He concluded that these effects vary little as between districts receiving large differences in rainfall, and they can thus be attributed mainly to the diminution of mean temperature with increasing altitude. Pearsall, in his conclusion, may have overlooked a number of other environmental factors which could possibly affect the growth of the moor-rush, such as soil mineral status, surface drainage or the intensity of grazing.

Data relating to the estimated yield of Sitka spruce in Taliesin Forest, north Cardiganshire, were made available to the author by the Forestry Commission in Wales. The choice of Sitka spruce from the variety of trees growing within the forest was determined by the extent, and altitudinal range of occurrence, of the species. Estimates of yield were relatively crude, being in yield-classes for each compartment or subcompartment of the forest, following the conventions of the Forestry Commission. A statistical relationship between estimated yield and approximate altitude of each compartment was then determined using a linear regression technique, the results of which are presented in visual form in Fig 2.5. A coefficient of linear correlation of -0.656, significant at the 0.1 per cent level, infers that forest productivity may be related to altitude. The problem remains to determine the cause of such a relationship.

Climate is a major determinant of forest productivity. The degree of exposure to strong winds, for example, particularly at higher altitudes, is a primary consideration in assessing the suitability of a site for afforestation. Decreases in forest productivity with increases in altitude may not necessarily be related to climate, because factors of soil type, vegetation and slope operate alongside climate in the determination of

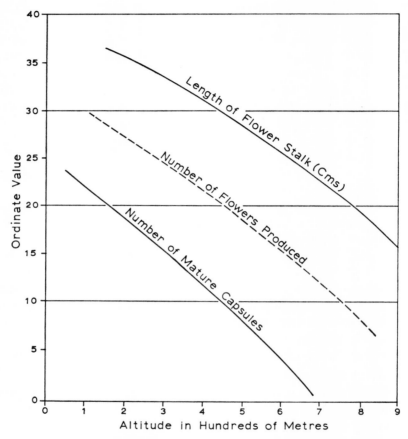

Fig 2.4 Effect of altitude on the growth of the moor-rush (*Juncus squarrosus*) (after W. H. Pearsall)

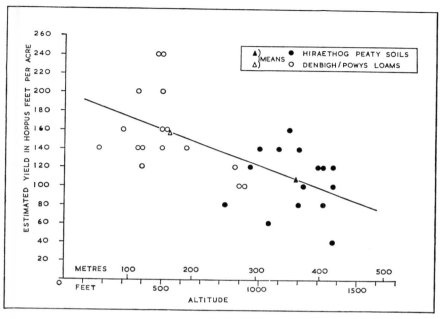

Fig 2.5 Relationship between the yield of Sitka spruce (*Picea sitchensis*) and altitude in Taliesin Forest, north Cardiganshire

the relative success of afforestation. The interpretation of the relationship between altitude and yield will, therefore, be confused by the proliferation of other factors involved.

To assess the possibility of variation in yields of Sitka spruce in Taliesin Forest, which may be due to factors other than climate, a simple division of the soils was made. The ill-drained peats and peaty podzols of the '*Hiraethog*' series (Rudeforth, 1970), with deep layers of organic material, were separated from the better-drained loams of the '*Denbigh*' soil series which has little surface accumulation of organic material. This resulted in two quite distinct distributions. It emerged that the line representing the relationship between altitude and estimated yield of Sitka spruce (Fig 2.5) could be replicated by joining the spatial means of the two soil-based distributions, and that, within the individual soil-based distributions, the relationship ceased to be significant. The analysis, despite the crudeness of the data and the elementary nature of its treatment, shows that any attempt to express a decrease in the yield of Sitka spruce with an increase in altitude, in terms of a climatically-defined growth gradient, is liable to overlook the more significant pedological variations.

These examples illustrate the need to place the role of climate into correct environmental perspective. It is easy to attribute altitudinal changes in plant growth, for example, to a shorter, thermally-defined growing season, or to fewer day-degrees of accumulated temperature, or increases in wind speeds, but clearly confusion can arise from such over-simplification.

The Final Assessment
The problems of assessing the significance of the changes in climate or climatic resources which occur as altitude increases are related (a) to the inadequacy of climatic data and sampling mechanisms within complex spatial and temporal variations, and (b) to the question of the viability of the final product in the context of marked changes in the total plant environment. The solutions lie in the extension of climatological surveys to cover the full altitudinal and topographical range of variation, and the determination of the exact role of climate and climatic factors within the plant environment.

Acknowledgement
This chapter presents a preliminary report on research conducted by the author, under the supervision of the editor, during his tenure of a NERC studentship at the Geography Department, University College of Wales, Aberystwyth.

References

BIRSE, E. L. *Assessment of Climatic Conditions in Scotland*, 3. *The Bioclimatic Sub-Regions*, Map and Explanatory Pamphlet, The Macaulay Institute for Soil Research (Aberdeen, 1971).

BUCHAN, A. et al. 'The Ben Nevis Observations', *Trans Roy Soc Edinburgh* (1905–10), 42–4.

GEIGER, R. *The Climate Near the Ground* (Harvard, 1965; 2nd edn).

HESS, M. 'A New Method of Determining Climatic Conditions in Mountain Regions', *Geographia Polonica*, **13** (1968), 57–77.

MANLEY, G. 'Meteorological Observations on Dun Fell', *QJ Roy Met Soc*, **68** (1942), 151–65.

——. 'Further Climatological Averages for the Northern Pennines', *QJ Roy Met Soc*, **69** (1943), 257–60.

——. 'The Effective Rate of Altitudinal Change in Temperate Atlantic Climates', *Geog Rev*, **35** (1945), 408–17.

MILLER, A. A. *Climatology* (1931).

OLIVER, J. 'Upland Climates in South Wales', in Taylor, J. A. (ed), *Hill Climates and Land Usage with Special Reference to the Highland Zone of Britain*, Memorandum No 3 (Aberystwyth, 1960).

PEARSALL, W. H. *Mountains and Moorlands* (1950).

RUDEFORTH, C. C. *The Soils of North Cardiganshire* (Harpenden, 1970).

SHELLARD, H. C. 'Averages of Accumulated Temperature and Standard Deviation of Monthly Mean Temperature over Britain', *MO Prof Notes*, No 125 (1959).

SMITH, L. P. 'Variations of Mean Air Temperature and Hours of Sunshine on the Weather Slope of a Hill', *Met Mag*, **79** (1950), 231–3.

——. 'Variations in Air Temperature and Humidity on the Weather Slope of a Coastal Hill', *Met Mag*, **81** (1952), 102–4.

TAYLOR, J. A. (ed). *Hill Climates and Land Usage with Special Reference to Highland Britain*, Geography Department, UCW, Memo No 3 (Aberystwyth, 1960).

——. 'Economic and Ecological Productivity under British Conditions', in *Weather and Agriculture* (Oxford, 1967), 137–45.

CHAPTER 3

M. B. ALCOCK, G. HARVEY and J. THOMAS

Measurement, Evaluation and Management of Climatic Resources for Grassland Production in Hill and Lowland Areas

Introduction

Agriculture is primarily dependent on photosynthetic systems. In this context, climatic resources can be said to include those factors which together determine the rate and duration of net photosynthesis in the field and hence the biological or total dry-matter yield of the crop. Biological yield may also be influenced by the way in which climate modifies the partition of dry matter in the plant which, in turn, also determines for some crops the economic yield or proportion of usable dry matter.

The *measurement* of climatic resources must involve the two basic questions. What are the appropriate climatic elements to be measured and how are these elements to be measured? The *evaluation* of climatic resources implies a predictable relationship between these resources and yield and requires a suitable mathematical framework upon which a theory can be developed and calculations performed. The *management* of climatic resources implies that, initially, these can be measured and evaluated and that management strategies can be developed and implemented.

In this paper all three statements will be discussed with particular reference to grassland production on specific lowland and hill areas of North Wales.

Measurement of Climatic Resources

The number of climatic factors operating in the atmosphere and the soil is large, and in practice it is often necessary or even logical to exercise some selection over what is to be measured. Determination of relevance depends on the biological response under consideration. A plant may respond to climate in the following ways:

1 Normal growth
   (a) to grow at varying rates,
   (b) to cease growth but also to survive.
2 Catastrophic growth
   (a) senescence of part of the shoot or root system but also to survive,
   (b) to die.

The necessary type of climatic measurement required to predict catastrophic growth or death may be different from that required to predict normal growth responses.

Further discussion of the measurement of any particular climatic element involves a consideration of frequency—whether continuous, discrete or integrated—and of sensitivity. Frequency depends on the speed of biological response to change in the climate element and on the extent to which that change affects total growth. Photosynthesis responds instantly to change in light flux density but for most purposes, hourly or even daily integrated readings are adequate. For some purposes hourly values of air and soil temperature can be estimated from measurements of daily maximum and minimum values using sine functions whereas, for others, actual hourly (or more frequent) measurements are necessary.

The degree of sensitivity required in measurement depends on the range of variation experienced in the field and on the magnitude of the biological response. In a special sense, sensitivity is an important issue when the measurement is to be used to calculate other physical processes.

Generally in research investigations it is desirable that: (a) measurements are kept to a minimum and are performed with reliable instruments; (b) a system is adopted which involves automatic recording combined with a computer-processing capability.

The basis of climatic measurement must be the measurement of the mesoclimate in a standard manner common to a network of climatic stations. In turn, however, the evaluation of such data must involve consideration of the mesoclimate/microclimate relationship as influenced by crop species and canopy development. Thus supplementary measurements are required at investigation centres which will involve specialised, sensitive instruments for monitoring the microclimate.

The primary factors linking climate with plant growth are (Idso (1968)):

1 Light flux density.
2 Carbon dioxide concentration.
3 Plant temperature.
4 Plant water potential.

These are factors having a direct effect on growth and other climatic elements such as humidity, wind speed and rainfall influence growth only indirectly. Day length is also important since developmental processes such as flowering are influenced by photoperiod. Soil oxygen concentration can also be significant.

1 Light Flux Density

Total daily and diurnal input of short-wave solar radiation income varies with latitude and time of year. It is also affected by cloudiness which accounts for a marked drop in daily values with increase in altitude in western Britain. The effects of slope and aspect are particularly well documented (Geiger (1965)) and these account for considerable differences which occur in hilly and mountainous terrain. The fraction of total short-wave radiation within the photosynthetically active spectrum (0.40–0.72μ) varies mainly through variation in atmospheric pollution, being as high as 50 per cent when the predominant air-flow is from polar regions to as low as 30 per cent when the wind flow is from the continent (Monteith (1972)). Above-crop measurements of short-wave radiation income are made with thermopile solarimeters and a convenient recording interval is fifteen minutes, the radiation being integrated over that period. Attenuation of light in crop canopies can be measured with tube solarimeters or it can be predicted if suitable measurements of canopy size and geometry are known—see Monteith (1965), de Wit (1965), Duncan *et al* (1967).

2 Carbon Dioxide Concentration

This may only rarely be a limiting factor in crop growth in Britain and average carbon dioxide concentrations in the range 310–20ppm are normally assumed. The concentration of carbon dioxide at the leaf surface is determined by the rate of photosynthesis and respiration within limits dependent on (a) carbon dioxide concentration above the canopy, (b) carbon dioxide evolution from the soil and (c) on diffusive resistances. Thus the transfer of carbon dioxide between air and sites of photosynthesis in the leaf can be expressed as:

$$P = \frac{pc\,(C_1 - C)}{r_a + r_s + r_m}$$

Equation 1 (Denmead (1968b))

where:
- P = rate of CO_2 exchange per unit area of leaf surface
- pc = density of CO_2
- C_1 = volumetric concentration of CO_2 at the photosynthesising site
- C = CO_2 concentration in the air
- r_a = external air resistance
- r_s = stomatal resistance
- r_m = mesophyll resistance

In general, changes in evolution of CO_2 from the soil are met with decreases in downward transfer of CO_2 from the air. Estimates of CO_2 concentration indicate the necessity for combinations of high light flux densities and low turbulent transfer before CO_2 concentrations could become limiting in this country (Monteith and Szeicz (1960)).

3 Plant Temperature

Direct measurements of plant temperature are difficult to make and most measurements are of correlated temperatures. With grasses, soil temperature, at least in the early months of the year, is appropriate. Temperatures taken at a depth of 10cm have proved useful in studies of grass growth (Alcock et al (1968)) but more recent measurements by Peacock (1971) indicate that a measurement depth of 2cm is more realistic.

Air temperature may become more relevant as the canopy develops. Leaf temperature may be lower or higher than air temperature in the canopy to an extent which can be predicted from *Equation 2*.

$$t_1 - t = \frac{Rn\ (r_a + r_1)/pL - \sigma}{Cp\ (r_a + r_1)/r_aL + \Delta} \quad \text{Equation 2 (Denmead (1968a))}$$

where:
- t_1 = temperature of the leaf surface
- t = temperature of the air
- Rn = net radiation
- r_a = air resistance to heat exchange
- r_1 = leaf resistance (mainly a stomatal resistance)
- p = density of air
- L = latent heat of evaporation
- σ = specific humidity deficit (q sat $(t) - q$ where q = specific humidity)
- Cp = specific heat of air at constant pressure
- Δ = slope of the saturation specific humidity versus temperature curve

Normally, a grass leaf well supplied with water will be fractionally below air temperature but, as the water supply diminishes and under high evaporative demand, leaf temperature can rise above air temperature.

4 Plant Water Potential

The rate of transpiration depends on a drop in water potential across the soil/plant/air water continuum and on the resistances to transpiration flux, thus:

$$F = \frac{\Delta \psi \eta}{r_n} \qquad \textit{Equation 3}$$

where:

- F = transpirational flux
- $\Delta \psi \eta$ = drop in water potential between the water source (soil) and sink (air)
- r_n = the sum total of resistances to water flux in soil, root, xylem, leaf (cuticle and stomata), boundary layer and external air.

The plant water potential—which exhibits a within-plant gradient—depends dynamically on (a) the evaporative demand of the air, (b) the availability of water in the soil and (c) the regulation of variable plant resistances such as the stomatal resistance.

A complete evaluation of these water relations requires measurements of radiation, humidity, air temperature, wind speed and rainfall, together with biological and physical measurements of plant and soil. The prediction of changes in water potential are difficult (Rijtema (1965)) and consequently, a simplified approach is often adopted. This depends on the prediction of soil moisture deficits from estimations of potential evapotranspiration and measurements of rainfall. A combination of current potential evapotranspiration and soil moisture deficit provides an index of current plant-water stress and can be used in predicting growth response. In this case plant-water stress is inversely proportional to the estimated attainment of potential evapotranspiration (Flinn (1971)).

Evaluation of Climatic Resources

Evaluation of climatic resources depends on an ability to predict yield from suitable measurements of climate. In the first instance the potential of the climate can be evaluated with an assumption being made that mineral nutrients, pests and diseases are not limiting plant growth. In classical growth-analysis yield can be expressed as the summation of the daily rates of dry-matter production which in turn is dependent on the net rate of photosynthesis per unit leaf area (net assimilation rate) and on the size of leaf area (leaf area index). The net rate of photosynthesis is the balance between gross photosynthesis and respiration. Gross photosynthesis is affected by light flux density and carbon dioxide concentration at the leaf surface, and by leaf water potential via its effect on stomatal resistances. Respiration is influenced by temperature and by light flux density in plants possessing photorespiration. Leaf growth is very sensitive to variation in temperature and water potential.

The prediction of dry-matter yield from climatic measurements depends, therefore, on the successful formulation of the mechanisms outlined above into an integrative model.

Early approaches to this problem were based on largely empirical, statistical correlations between yield and a few major limiting climatic elements such as temperature or water availability (Wang (1963)). These statistical models are characterised by their specificity to the crop and region and by their limited resolution in terms of accounting for minor changes in climate.

In more recent times attention has been directed to considering the crop as a photosynthetic system and various discrete mathematical models have been developed for estimating potential canopy photosynthesis and annual dry-matter production from measurements of short-wave radiation—see Monteith (1965), de Wit (1965), Duncan *et al* (1967). Thus, estimates of potential production of an area can be made for appropriate crop species.

The availability of water is frequently a limiting factor in crop production and consequently there is a need to account for this in predicting the yield of dry matter. Penman (1971) has developed a yield equation which is based on an assumed efficiency of conversion of solar radiation to plant dry matter and on an estimation of 'effective transpiration'. Thus:

$$Y = K\,Ea \qquad \text{Equation 4}$$

where:
- Y = yield of dry matter
- K = constant = maximum possible response to irrigation
- Ea = effective value of potential transpiration contributing to plant growth

in which:

$$K = \frac{\Delta Y}{I} = 39\ \xi\ t\ ha^{-1}\ cm^{-1} \qquad \text{Equation 5}$$

$$Ea = E_T + D_1 - D_m \qquad \text{Equation 6}$$

where:
- ξ = efficiency of conversion of solar radiation to plant dry matter
- I = irrigation
- E_T = potential transpiration from a turf surface
- D_1 = limiting soil moisture deficit
- D_m = maximum soil moisture deficit

The values of ξ and D_1 are chosen for specific crop situations.

Rijtema (1970) has developed a model which relies more on interpretating physiological responses in the crop and as a consequence is probably more flexible and of higher resolution. The basic equation is:

$$P = a\,\frac{4 \cdot 9}{r_a + r_s + r_m}\,S_c P_{pot} \qquad \text{Equation 7}$$

where:
- P = rate of dry-matter production of the actual crop
- a = a factor correcting for respiratory losses
- r_a = the exchange resistance between the bulk air and the effective surface canopy (sec^{-1} cm^{-1}) based on measurements of wind speed and crop height

r_s = diffusion resistance of the crop (stomatal resistance—sec^{-1} cm^{-1}) based on estimates of transpiration and soil water content

r_m = mesophyll resistance of the crop (sec^{-1} cm^{-1})

4·9 = sum of the resistances for the standard crop

S_c = fraction of the surface covered by the plants—modified on a basis of an empirical relationship with soil water content

P_{pot} = potential production of a standard crop which is dependent on radiation income as the main determining factor

The application of both of these models is restricted to the growing season which can be loosely defined as the period when mean daily temperatures exceed 6° C. While Rijtema's model could allow for variation in S_c (LAI) on a basis of temperature no such use has been reported and indeed there are real difficulties in introducing temperature dependency for leaf growth in discrete models.

A more complete evaluation of climate resources and accurate prediction of yield is more likely to be achieved by the use of dynamic models of crop growth which take into account the various physiological mechanisms determining the relationships between crop growth and climate. Progress with this approach has been made in recent years by de Wit and Brouwer (1968, 1969) and Brouwer and de Wit (1969) with maize (*Zea mays*), and Patefield and Austin (1971) with red beet (*Beta vulgaris* L), and Fick (1971) with sugar beet (*B. vulgaris* L).

In this paper we describe preliminary steps in developing a dynamic model of grass growth.

In a paper to a symposium on hill farming by Alcock and Lovett (1968), we reported that differences in pasture production on lowland (6m OD) and hill (290m OD) sites from various grass species were mainly confined to the spring period. Under conditions of non-limiting nutrients, it was concluded that this difference in production was a result of lower temperatures on the hill limiting leaf growth, and that ultimately it would be possible to predict growth in the spring from measurements of temperature and short-wave radiation.

Development of a Model of Spring Growth

The initial assumptions in constructing a model of grass growth were: (a) the period was restricted to primary growth in spring, (b) mineral nutrients and water were not limiting, (c) loss of dry matter through senescence and decay was negligible and (d) photosynthesis was not limited by the rate of utilisation.

The model could be divided into two parts:
1 the development of leaf area (LAI) assuming that the supply of substrate did not limit growth;
2 the prediction of rates of dry-matter production of the aerial parts of the plant (shoots).

Dry-matter production was to be predicted for a S.23 perennial ryegrass (*Lolium perenne*) sward growing at 6m OD and at 290m OD.

1 Leaf-area Index

The relationship between the relative leaf growth-rate (RLGR $cm^2/cm^2/$day) and mean daily air temperature recorded at ground level in a Stevenson screen was determined for periods of minimal water stress in a separate experiment. For prediction of increase in LAI an initial value was taken from field measurements.

Increase of LAI in the model was determined by the mean daily air temperature. The appropriate RLGR at an observed mean temperature was used to calculate increase of LAI ($=$ LAI $\times$ RLGR$_t$), and the crop LAI was updated at the end of each day. The relationship between observed and predicted values of LAI is shown in Fig 3.1. In 1969 the soil was near field capacity during the spring period but in 1970 an appreciable soil moisture deficit developed in late spring and the model, according to expectation, resulted in an over-estimate of LAI.

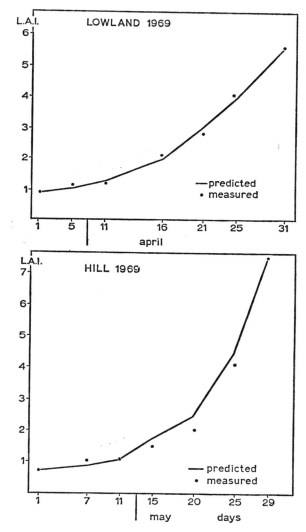

Fig 3.1 Leaf-area index of perennial rye-grass on hill and lowland sites in North Wales

2 Dry-matter Production

(a) CANOPY PHOTOSYNTHESIS was calculated using the model described by Monteith (1965) with coefficients suitable for S.23 perennial ryegrass. The choice of coefficients was based on evidence from a literature search and accounted for a prostrate type canopy and an average leaf photosynthesis/light flux density relationship.

Light penetration was based on:

$$I(L) = [0 \cdot 4 + (1 - 0 \cdot 4) 0 \cdot 10]^L I(0) \qquad \text{Equation 8}$$

where:
- I = light flux density
- L = leaf-area index
- $0 \cdot 4$ = a constant which specifies the average orientation and arrangement of leaves of a canopy layer of unit LAI
- $0 \cdot 10$ = mean transmission of visible light through a leaf

The rate of photosynthesis–light flux density relationship was based on:

$$P(I) = \left(0 \cdot 88 + \frac{0 \cdot 05}{I}\right)^{-1} \text{g } CH_2O/m^2/\text{leaf area/hr} \qquad \text{Equation 9}$$

where:
- I = cal cm^{-2} min^{-1}
- P = rate of photosynthesis

(b) PARTITION OF ASSIMILATES This crucial process was based on a distribution function determined from actual field measurements.

It was calculated from the ratio of $\dfrac{(dW_S)}{(dt)}$ to $\dfrac{(dW_T)}{(dt)}$

where:
- W = dry weight
- S = shoot
- T = total weight
- t = time

CLIMATIC RESOURCES IN HILL AND LOWLAND AREAS 77

It was utilised as a time dependent function of the proportion of assimilates entering the shoot. Actual values ranged from 0·1 to 1·0.
(c) RESPIRATION Respiration of shoots only was considered. This was based on the work of McCree (1970) and used the equation:

$$R = 0\cdot25\,P + 0\cdot015\,W \qquad \text{Equation 10}$$

where:
- R = rate of respiration of the whole plant
- P = rate of gross photosynthesis of the whole plant
- W = dry weight of living material on the plant

Shoot growth = Canopy photosynthesis × Shoot distribution function
— Shoot respiration

The predicted and observed rates of growth of S.23 perennial rye-grass swards in hill and lowland environments are shown in Fig 3.2. The agreement is close and indicates that differences in growth rate between the sites could be predicted on the basis of differences in LAI

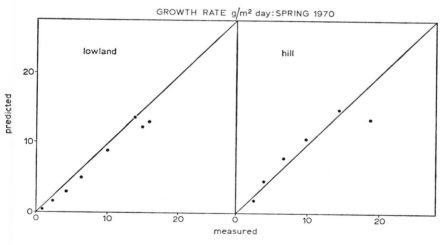

Fig 3.2 Rates of growth of perennial rye-grass on lowland and hill sites in North Wales

and solar radiation. In a year with lack of water stress increase in LAI could be predicted and growth curves of dry weight could also be simulated.

Further investigations are required before a model of wide generality and high resolution can be formulated. Items receiving attention are:

1 The prediction of the onset of net increase in LAI.
2 The use of a more suitable temperature measurement for predicting increase in LAI—such as 2cm soil temperature—Peacock (1971).
3 The prediction of plant water potential and the relationship of this to increase in LAI.
4 The prediction of dry-matter partition to avoid the use of an empirical distribution function.

At present predictions are based on daily values, ie mean daily temperature is calculated from continuous records whilst solar radiation is assumed to vary sinusoidally between sunrise and sunset and is estimated from daily totals of radiation. As water-stress factors are introduced, predictions may have to be based on shorter time-steps, possibly of the order of 1 hour. When growth at other times of the year is considered then additional plant factors will have to be considered such as the population density of meristems and dry-matter partition between leaves, stems, and dead material in the shoot.

The Management of Climatic Resources

A dynamic model of growth can be used to evaluate: (a) the potential of a given area for grassland production and (b) the value of alternative strategies of management or exploitation of climatic resources such as:

1 Plant breeding—eg to assess quantitatively the value of a high capacity for rapid leaf growth at low temperature (assuming high winter survival).
2 Exploitation of local climate—eg to assess the value of grazing strategies based on differences in growth due to variation in altitude and aspect.
3 Modification of pasture microclimate—eg irrigation, shelter.

At present the effect of shelter on pasture growth is being investigated and the following discussion is concerned with direct observations of the effect of shelter on the growth of pasture in the spring. This is a necessary first step before developing a growth model which will also incorporate the prediction of the effect of shelter on pasture microclimate.

Shelter changes the pasture microclimate primarily by its effects on turbulent transfer. The main consequences are a reduction in evaporative demand which when soil water is not limiting will result in a reduction of transpiration. Plant water potential is influenced by the interactions of water supply (soil water) and evaporative demand, and, generally, shelter has the effect of decreasing the negative values of water potential in the plant. Theoretically, this should lead, firstly, to increased relative leaf growth rates and, secondly, to increased rates of photosynthesis as a result of decreased stomatal resistance. Air and soil temperatures may be increased during the day and reduced during the night. These variations in temperature would be expected to influence the relative leaf growth rate and, to a lesser extent, photosynthesis.

The Effect of Shelter on the Growth of S.23 Perennial Rye-grass

In 1969—which had a wet spring—no effects of shelter on the growth of grass were observed which suggested that temperature differences are negligible. In 1970, however, the spring was dry with soil moisture deficits developing in late April and May. Fig 3.3 shows the effect of shelter on grass growth measured at a distance of 4h from a NETLON shelter barrier.

Curves were fitted to observations of dry weight made every 3–4 days and were of the form:

$$\log_e W = a + b_1 t + b_2 t^2 + b_2 t^3$$

where:

W = dry weight, either of shoots or of shoots plus roots
t = time
b_1, b_2, b_3 are regression coefficients

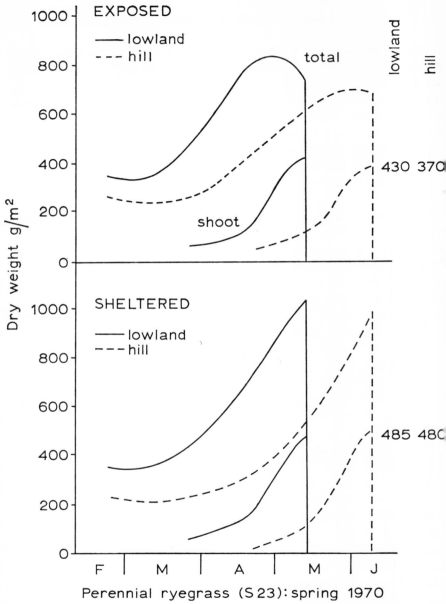

Fig 3.3 Effect of shelter and altitude on the growth of perennial rye-grass in spring in North Wales

Shelter increased dry-matter production, the increase in shoot weight being 12 per cent on the lowland and 30 per cent on the hill. The difference in response between lowland and hill can be explained by the observed relationship between the crop growth rate (g/m²/day) and

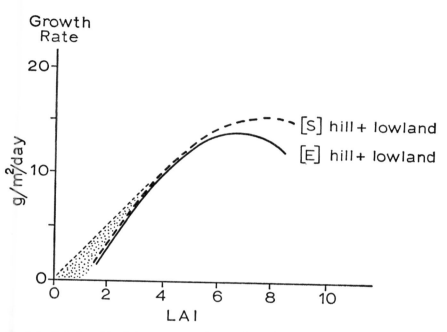

Fig 3.4 Relationship between rate of growth and leaf-area index of perennial rye-grass

LAI (see Fig 3.4). If it is accepted that shelter affected growth primarily by its influence on leaf expansion (as Fig 4.4 suggests) and that this occurred at the onset of water stress, then this would be important only at the later stages of canopy development on the lowland, whereas on the hill water stress occurred at a much earlier stage of development.

F

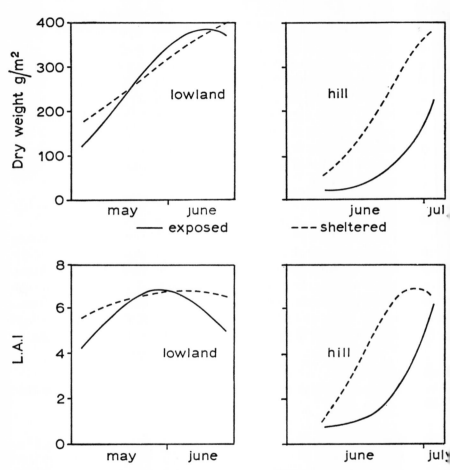

Fig 3.5 Effect of shelter and altitude on the growth of white clover in spring in North Wales

Since increases in growth rate only occur with increase in LAI up to about 6–7, the beneficial effects of shelter would clearly be greater on the hill.

The Effect of Shelter on the Growth of S.184 White Clover
The effect of shelter on S.184 white clover is shown in Fig 3.5. On the lowland, LAI was high at the beginning of the season, soon reaching a ceiling value, and shelter had little effect on either LAI or dry weight. On the hill, LAI was very low at the beginning of the season and there were large areas of bare soil. Soil heating effects in shelter (in a wet spring) could account for most of the rapid increase in LAI and dry weight. In both sheltered and exposed plots the LAI ultimately reached the same ceiling value.

Conclusion
In this paper attention has been drawn to the primary factors linking climate and plant growth and the problems of making suitable climatic measurements. Emphasis is placed on the development of dynamic growth models for evaluating climatic resources. The output from such models could provide the first basic step in an ultimate attempt to assess the economic value of such resources. Some preliminary observations of the effect of modifying pasture microclimate by shelter are made.

Acknowledgement
We wish to acknowledge the considerable help of E. Lloyd-Jones in field work.

References
ALCOCK, M. B. and LOVETT, J. V. 'Analysis of Environmental Influence on Productivity', *Proc Symp Hill Land Product*, Occ Symp No 4 (1968), British Grassland Society, 20–9.
ALCOCK, M. B., LOVETT, J. V. and MACHIN, D. 'Techniques Used in the Study of the Influence of Environment on Primary Pasture Production

in Hill and Lowland Habitats', in Wadsworth, R. M. (ed), *The Measurement of Environmental Factors in Terrestrial Ecology*, Symp British Ecological Society (Oxford, 1968), 191–203.

BROUWER, R. and DE WIT, C. T. 'A Simulation Model of Plant Growth with Special Attention to Root Growth and Its Consequences', in Whittington, W. J. (ed), *Root Growth* (1969), 224–44.

DENMEAD, O. T. 'The Energy Balance of Plant Communities', *Proc WMO Seminar Agr Meteorol*, 1966, Bureau of Meteorology, Australia (Melbourne, 1968a), 71–105.

——. 'Carbon Dioxide Exchange in the Field: Its Measurement and Interpretation', *Ibid* (1968b), 445–82.

DUNCAN, W. G., LOOMIS, R. S., WILLIAMS, W. A. and HANAU, R. 'A Model for Simulating Photosynthesis in Plant Communities', *Hilgardia*, **38** (1967), 181–205.

FICK, G. W. *Analysis and Simulation of the Growth of Sugar Beet (Beta vulgaris* L), PhD thesis (University of California, Davis, 1971).

FLINN, J. C. 'The Simulation of Crop Irrigation Systems', in Dent, J. B. and Anderson, J. R., *Systems Analysis in Agricultural Management* (Sydney, 1971), 123–51.

GEIGER, R. *The Climate Near the Ground* (Harvard, 1950; 4th edn 1965).

IDSO, S. B. 'A Holocoenotic Analysis of Environment-Plant Relationships', *Tech Bull* **264** *Agr Exp Stat* (Univ Minnesota, 1968).

MCCREE, K. J. 'An Equation for the Rate of Respiration of White Clover Plants Grown under Controlled Conditions', in Setlik (ed), *Prediction and Measurement of Photosynthetic Productivity* (1970), Proc IBP/PP Tech Meeting Trebon, 1969.

MONTEITH, J. L. 'Light Distribution and Photosynthesis in Field Crops', *Ann Bot*, **29** (1965), 17–37.

MONTEITH, J. L. and SZEICZ, G. 'The Carbon Dioxide Flux over a Field of Sugar Beet', *QJ Roy Met Soc*, **86** (1960), 205–14.

——. *Personal Communication* (1972).

PATEFIELD, W. M. and AUSTIN, R. B. 'A Model for the Simulation of the Growth of *Beta vulgaris* L', *Ann Bot*, **35** (1971), 1227–50.

PEACOCK, J. M. 'Plant and Crop Interaction with the Environment', *Ann Rep Grassld Res Inst, 1970* (1971), 55–7.

PENMAN, H. L. 'Water as a Factor in Productivity', in Wareing, P. F. and Cooper, J. P. (eds), *Potential Crop Production* (1971), 189–99.

RIJTEMA, P. E. 'An Analysis of Actual Evapotranspiration', *Agr Res Rept* (Wageningen, 1965), 659.

———. 'The Effect of Light and Water Potential on Dry Matter Production of Field Crops', *Symp Plant Responses to Climatic Factors*, UNESCO (Uppsala, 1970).

WANG JEN-YU. *Agricultural Meteorology* (Milwaukee, 1963).

WIT, C. T. DE. 'Photosynthesis in Leaf Canopies', *Agr Res Rept* (Wageningen, 1965), 663.

WIT, C. T. DE and BROUWER, R. 'Über ein Dynamisches Modell des Vegetativen Wachstum von Pflanzenbestanden', *Zeit für Angewandte Bot*, **42** (1968), 1–12.

———. 'The Simulation of Photosynthetic Systems', in Setlik (ed), *Prediction and Measurement of Photosynthetic Productivity* (1970), Proc IBP/PP Tech Meeting, Trebon, 1969, 16–32.

CHAPTER 4 J. M. PEACOCK
and J. E. SHEEHY

The Measurement and Utilisation of Some Climatic Resources in Agriculture

Introduction

The use of plants in agriculture represents a major exploitation of climatic resources. The growth of green plants depends on photosynthesis, and every year this process results in the fixation of 40,000 million tonnes of carbon of which 16,000 million tonnes are fixed on land. With the annual world production of coal and oil only reaching approximately 2,000 million and 1,000 million tonnes respectively, agriculture emerges as the principal extractive industry (Costes (1964)). The energy exchanges involved in the process of fixation of carbon by photosynthesis are immense, Rabinowitch and Govindjee (1969) estimated that 3×10^{17}K calories are stored by land plants annually. However, the overall efficiency of conversion of photosynthetically active radiation into plant material is only about 1·2 per cent and consequently the economic significance of research aimed at improving the efficiency of this process is clear.

Representatives of the *Gramineae*, wheat, rice, sugar cane, maize and pasture grasses, provide 60 per cent of the world's food. They occupy 46 million square kilometres—24 per cent of the earth's vegetative cover (Barnard (1964))—and in Britain alone pasture grasses occupy 65 per cent of the agricultural land (Woodford (1969)). The most widely sown species in British agriculture is *Lolium perenne* and it is a cultivar S24 of this species that we have selected to investigate how a crop might achieve a better utilisation of the seasonal input of solar energy.

If water and soil nutrients are not limiting, the annual dry-matter production of a perennial rye-grass crop grown in Britain may be as high as 20,000kg ha^{-1} (Cooper and Breese (1971)). Nevertheless, the efficiency of light energy conversion is very low, approximately 2 per cent; this efficiency is largely determined by the size, duration and structure of the photosynthetic surface of the crop (Watson (1971)). The first two of these factors are known to be temperature dependent; the first part of the chapter is concerned with an investigation of the influence of temperature on the rate of leaf expansion. The second part of the chapter studies the influence of the size and geometrical arrangement of the photosynthetic surface on the utilisation of light energy.

The Problems of Examining the Effect of Temperature on Crop Growth in the Field
There is considerable variation in the response to temperature of cultivars and ecotypes of *Lolium*, *Dactylis* and *Festuca*—Cooper (1964), Robson (1967). In particular, certain races can maintain leaf growth at relatively low temperatures. The general effects of temperature on plant growth are well documented—Went (1957), Leopold (1964), and have been comprehensively reviewed by Langridge and McWilliam (1967). However, most early work was on spaced plants in controlled environments and few attempts were made to assess how temperature affects the growth of crops in the field. Furthermore, these attempts have usually made use of mean air temperatures, or temperatures measured in a Stevenson screen (Bull (1968)), and have ignored, in general, possible differences between the effects on growth of soil temperature, air temperature, and the temperature of specific plant organs. Such effects have therefore been examined in our work where we have measured profiles of canopy and soil temperature. This was achieved by placing thermistors and platinum resistance thermometers at various depths within the soil and at intervals from ground level to the top of the crop. Aerial sensors were shielded from direct radiation with wire gauze shields painted white. Measured temperatures were accurate to $\pm 0.5°$ C

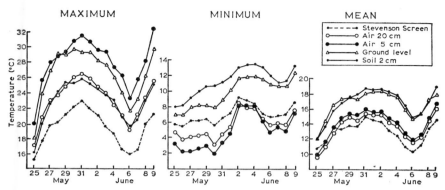

Fig 4.1 The comparison of temperatures measured in a Stevenson screen with maximum, minimum and mean temperatures measured at various levels within and beneath a crop of perennial rye-grass cv S24

and recorded by a data logger on punch tape. Fig 4.1 shows maximum, minimum and mean temperatures measured in the canopy of a perennial rye-grass crop and in the soil beneath it following a cut on 24 May, compared to those measured in a Stevenson screen. It is evident that screen temperatures were different from the soil and air temperatures inside the plant canopy.

Fig 4.2 shows instantaneous temperature profiles through a day and a night in a perennial rye-grass crop before and after cutting. This demonstrates the insulating properties of the crop which may have important consequences for plant growth (Kleinendorst (1972)). These figures also demonstrate that there is a diurnal variation in the thermal microclimate of the crop. However, it is difficult to use these instantaneous profiles in growth studies because of the problem of making any meaningful instantaneous measurements of growth. Temperature fluctuations can be measured instantaneously or over short periods of minutes or hours, whereas the usual measurement of crop growth, in terms of dry-matter production, is normally made at intervals of several days, and is the resultant of changes in its environment over this period.

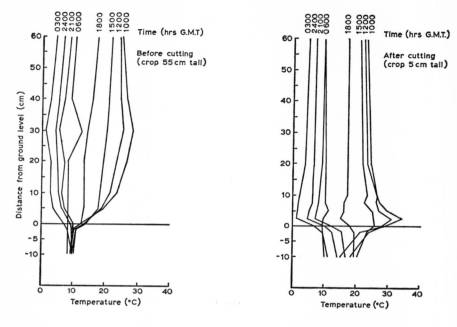

Fig 4.2 Temperature profiles within a grass crop: (a) shows a series of soil and air temperature profiles taken at intervals during a 24 hour period through a crop of perennial rye-grass 55cm tall; (b) shows a similar series of profiles taken during the same 24 hour period, in an adjacent area of the same crop cut four days before (from Peacock and Stiles, 1969)

An initial objective in this investigation was, therefore, to establish a more suitable measurement of growth.

The Effect of Temperature on Leaf Growth within the Grass Canopy

Temperature is known to affect the rate of leaf growth—see Mitchell (1953), Cooper (1964) and Robson (1969). Consequently, in this study an attempt was made to correlate the rate of leaf growth in the field with temperature at various levels within the crop and the soil. The rate

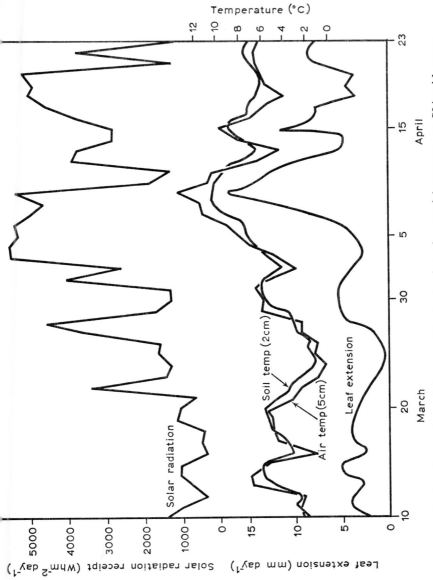

Fig 4.3 Comparison of the rate of leaf extension of perennial rye-grass, cv S24, with mean temperature and radiation receipt

of leaf growth was determined by measurements of the linear expansion of the youngest growing leaf in perennial rye-grass cv S24. Measurements were made daily on three sets of six tillers initially selected at random in the crop and the average rate of leaf extension (mm/day) was calculated. Fig 4.3 shows the rate of leaf extension, temperature and solar radiation during spring. These data show that leaf extension is very sensitive to changes in mean temperature, whereas there was no simple relation with radiation. When leaf extension was plotted against mean temperature measured at various levels in the soil and canopy and response curves were fitted to the data (Fig 4.4) they were found to be exponential and were described by the equation:

$$y = \alpha + \beta(\delta)^x$$

where:

y = leaf extension in mm day^{-1}
x = temperature in °C; α, β and δ are constants

The curves show that no distinction could be made between the effects of soil and air temperatures on leaf extension. However, Kleinendorst and Brouwer (1965) had suggested that root temperature controlled growth irrespective of air temperature. To investigate this problem further, an attempt was made to separate the effects of soil temperature and air temperature on growth using another method.

Soil Heating Experiment

Within the area of perennial rye-grass, the turf in two plots was lifted and four heating cables (Canham (1964)) were installed in one of the plots 17m × 2m before replacing the turf. The cables were laid on a metal frame at a depth of 15cm with adjacent loops 10cm apart, these provided up to 100watts/m².

Earlier work (Roy and Peacock (1972)) had shown that growth in the spring was noticeable when soil surface temperatures were approximately 5°C. Using the cables, soil temperatures were raised by 5°C in February, independently of air temperature. The rates of leaf expansion

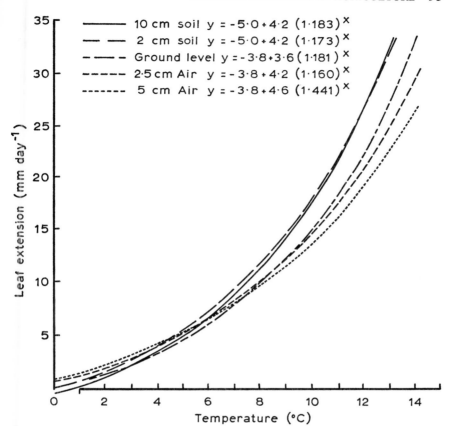

Fig 4.4 Response curves of leaf extension of perennial rye-grass, cv S24, plotted against mean temperatures measured at various levels in the soil and in the plant canopy

increased in the heated plots, suggesting that soil temperature influenced leaf growth. However, the upward heat flux from the soil resulting from the soil heating had a slight effect (1° C) on air temperatures which complicated the analysis of these observations. Nevertheless it is possible to show that there was a discrete zone in the soil where temperature was having significant effect on leaf extension. Fig 4.5 shows mean soil and air temperature profiles in the crop from different days during spring, but when the rate of leaf extension in heated and unheated plots were the same. Two sets of profiles for two widely different rates of leaf extension are shown. In both cases the profiles cross just below ground level where there was an identical temperature in both

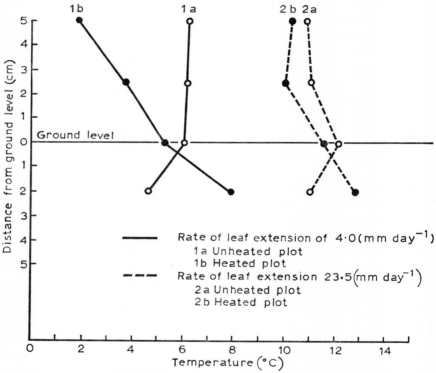

Fig 4.5 Profiles of mean soil and air temperatures in a crop of perennial rye-grass, cv S24, when the rates of leaf extension in heated and unheated plots were the same. Two sets of profiles are shown for two widely different rates during the spring period (from Peacock (1971))

the heated and unheated plots. Since the rate of leaf extension in both plots was the same, it appears that it is the temperature at this level in the profile which determines the rate of leaf extension. This is significant because in early spring, the growing point or shoot apex of perennial rye-grass is at, or just below, the soil surface (Peacock (1971)). Similar results were obtained by Kleinendorst and Brouwer (1970) and Watt (1971), working in controlled environment experiments with maize.

The effect of temperature on leaf extension during the rest of the year was also examined. During the summer period the effect of water stress complicated the results and only the data for the autumn period are discussed here. Results showed that temperature was again acting at the growing point. The response of the rate of leaf extension to temperature during this period was similar to the spring period in that the response was exponential and could be described by a curve similar to those shown in Fig 4.4 (Peacock (1972)).

We have shown that, during both spring and autumn, leaf growth of the crop is apparently restricted by the temperature of the growing points.

The Use of a Simple Model to Describe the Growth of a Grass Crop

Because the area above the heating cables was small it was not possible to take repeated destructive samples of the crop to determine its growth rate. We have, therefore, used a mathematical model to examine the influence of soil temperature on the efficiency of utilisation of solar energy.

The Monteith (1965) model was chosen because it is simple, and has been used to give reasonable predictions of crop growth rate—see Osman (1971), Patefield and Austin (1971), Sheehy and Cooper (1973). This model is based on the assumption that canopy structure can be described by a simple mathematical function. A value of this function, S, specifies the average orientation and arrangement of leaves of a canopy layer, of unit leaf-area index, and represents the amount of light passing through the layer without interception. If the values for leaf transmissivity (TR), and leaf-area index (L), are known, the distribution of light within the canopy can be estimated. Crop photosynthesis may then be calculated by combining equations describing the distribution of light within the canopy, the diurnal variation of light intensity, and the rate of gross photosynthesis of individual leaves as a function of light intensity. This function has the form:

$$Pg = 2 \int_0^{h/2} \left(\sum_{n=0,1} An \left[a + \frac{b}{I^* (1-S) TR^n \sin \pi th} \right]^{-1} \right) dt$$

where:
- Pg = the rate of gross canopy photosynthesis
- A = the area receiving light energy
- a = the reciprocal of the maximum rate of photosynthesis
- b is inversely proportional to the quantum efficiency in weak light
- t = time in minutes
- h = the total number of minutes for which the canopy is illuminated
- I^* = the apparent mid-day light intensity

The model takes no account of temperature although the rates of net photosynthesis and dark respiration of grass leaves are known to be strongly dependent on temperature—see Cooper and Tainton (1968), Beevers (1969), Charles-Edwards et al (1971). We have, therefore, modified the model to take account of the influence of temperature, over the ambient range experienced by our plots during the experimental period. Two assumptions are made. The first of these is that the maximum rate of gross photosynthesis P_m is linearly dependent on temperature, and can be described by an equation of the form:

$$P_m = g T$$

where:
- g = a constant
- T = temperature

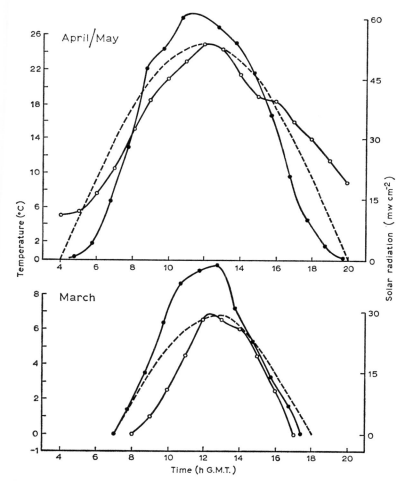

Fig 4.6 The diurnal variations of the mean measured and calculated mid-canopy temperatures, and the mean diurnal variations of short-wave radiation as measured by a Kipp solarimeter

G

The second assumption is that the average temperature within the canopy, T, varies sinusoidally through the day, and can be described by the function:
$$T = Tm \, \text{Sin} \, \pi \, t/h$$
where:
Tm = the maximum temperature
t = time
h = daylength

Fig 4.6 shows the mean diurnal variation of total incoming short-wave radiation (0·4–2·5μm), mean mid-canopy temperature, and the fitted sinusoidal temperature variation for the mid-canopy. It can be seen that the temperature tends to be over-estimated in the morning and sometimes under-estimated towards the end of the day. When these modifications are included in the model, predicted gross canopy photosynthesis is given by an expression of the form:

$$Pg = \sum_{n=0,1} An \, 2h(1-S)\pi^{-1} \left[\frac{(1-s)}{g \, Tm} + \frac{b}{I^* \, TR^n} \right]^{-1}$$

Crop growth rates (shoot and root) can then be obtained by allowing 45 per cent of gross photosynthesis for respiration (Robson, personal communication, and Ryle *et al* (1972)). A dry weight of 1g was assumed to be equivalent to 1g hexose and derived from 0·682g carbon dioxide.

Measured values of the various parameters (Table 4.1) were used to obtain estimates of the crop growth rate of the unheated plots, which were then compared with the measured values. The temperatures in the heated plots were lower than those in the control as a result of the greater insulation of the larger leaf-area index present in the heated plot. The agreement was good ($P<0.001$, Fig 4.7), so with some confidence we then estimated the crop growth of the heated plots. To do this we assumed that the area of leaf surface (L) and its geometrical arrangement (S) were identical for the heated and control plots when they were intercepting the same amount of light (Fig 4.8). The heated

Table 4.1

INPUTS OF THE MODEL, AND EFFICIENCY OF CONVERSION OF LIGHT ENERGY DERIVED FROM IT

Days (1971)	Radiation (0.4μm–2.4μm) Whm^{-2} day^{-1}	Day-length h	Maximum mid canopy temp °C (heated)	Maximum mid canopy temp °C (control)	L (calculated) (heated)	L (meas'd) (control)	I/I_0 (meas'd) (heated)	I/I_0 (meas'd) (control)	Efficiency of light energy conversion (%) (0.4μm–0.7μm) (heated)	(control)
11–16 Feb	1,195.5	9.80	9.38	9.38	1.53	1.22	0.85	0.87	0.92	0.78
16–23	1,946.9	10.25	11.10	11.60	1.72	1.35	0.84	0.87	0.94	0.77
23–1	1,955.6	10.71	7.69	8.69	1.93	1.42	0.75	0.86	1.32	0.82
1–9 March	2,202.7	11.19	7.10	7.50	2.75	1.50	0.57	0.85	1.84	0.78
9–16	2,228.3	11.68	12.07	12.27	3.86	1.77	0.41	0.83	2.98	1.01
16–23	2,341.1	12.11	11.48	11.89	4.50	1.85	0.32	0.78	3.30	1.19
23–30	2,694.7	12.62	12.80	13.18	5.15	1.90	0.25	0.71	3.58	1.53
30–5	2,003.8	13.03	11.50	11.68	5.75	2.29	0.19	0.63	4.05	2.01
5–13 April	2,948.2	13.49	12.50	13.01	6.29	3.05	0.14	0.53	3.86	2.39
13–20	3,621.6	15.95	13.48	13.87	6.70	4.00	0.10	0.38	3.96	3.06
20–26	2,590.0	14.39	12.50	13.08	7.05	5.92	0.08	0.22	4.21	3.99
26–4 May	6,028.3	14.80	13.02	14.13	7.92	6.73	0.06	0.11	3.14	3.21

Maximum rate of photosynthesis/unit L $(Pm) = 0.606$ mg dm^{-2} min^{-1} at 15°C
$b = 22.68 \times 10^4$ J/gm (0.4μm–2.4μm) } Sheehy (1970)
$TR = 0.288$
$I/I_0 =$ fractional light transmission
$g = 0.0404$ mg dm^{-2} min^{-1} °C^{-1}

$$S = \frac{(I/I_0)^{1/L} - TR}{1 - TR}$$

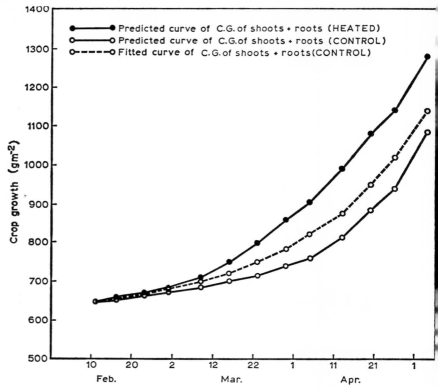

Fig 4.7 Crop growth of total shoot and root of perennial rye-grass

plots were found to have a considerably higher rate of predicted crop growth than the unheated controls (Fig 4.7); as might be expected from their higher rate of leaf elongation.

These rates of dry-matter production can be used to calculate the efficiency of conversion of light energy (0.4–$0.7\mu m$) into chemically bound energy, for both the heated and unheated plots, using a factor of 17,850J/g dry matter (Hunt (1966)). The results in Table 4.1 show

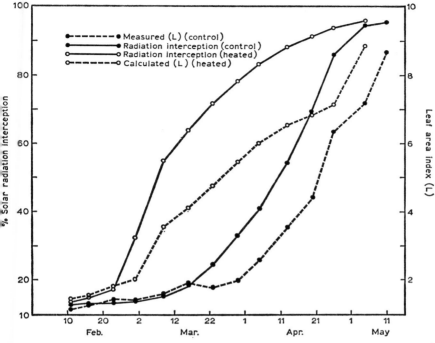

Fig 4.8 Interception of solar radiation and leaf-area index in heated and control plots of perennial rye-grass

that for much of the experiment the efficiency of the heated plot was greater than that of the control plot. This was due entirely to greater light interception resulting from the greater rate of leaf area expansion.

Our model can now be used in a speculative fashion to examine the effects of an increase in leaf-area index (L), either alone or when accompanied by a change in canopy structure (S), and also an increase in the maximum rate of photosynthesis per unit leaf area (P_m), which is equivalent to an increase in temperature in our model. Fig 4.9 shows the effect of altering these parameters. An increase in L has a great effect on the crop growth rate. Furthermore, this effect is enhanced in

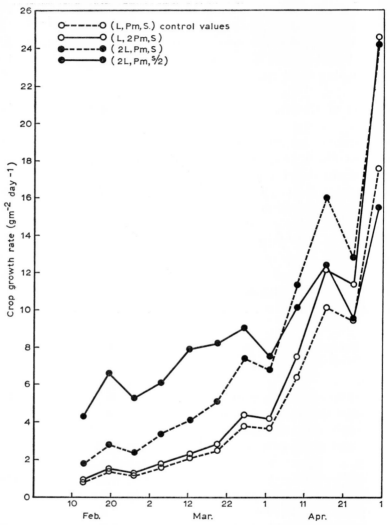

Fig 4.9 The effect of altering three parameters on predicted crop growth rates

early spring (10 February–4 April) when the increase in L is accompanied by a change in the angular structure $(S/2)$ so that there is a greater interception of the available light energy. An increase in P_m also exerts considerable influence, particularly at the later stages in the development of the crop.

The Effect of Genetic Variation on Factors Influencing Crop Growth Rate

The influence of genetic variation in crop structure and maximum rate of photosynthesis (P_m) on crop growth rate was examined in early June 1969 when the forage grass canopies were intercepting 95 per cent of the incoming light energy (Sheehy and Cooper (1973)). Two examples of grasses of each of three contrasting growth habits were grown in small plots without limitation of water or mineral nutrients (Fig 4.10). In addition to crop growth rate, measurements were also made of the photosynthetic rate of individual leaves attached to plants in the canopy, the percentage of leaves having inclinations to the horizontal greater than 60°, and components of the canopy structure which enable the extinction coefficient to be determined as defined by Beer's law. This states:

$$K_{vis} = \frac{1}{L} \log_e \left(\frac{I_o}{I}\right)$$

where:
K_{vis} = the extinction coefficient of visible radiation
L = the leaf-area index
I_o = the incident light energy
I = the light penetrating a leaf-area index L

The relationship between crop growth rate and P_m, the reciprocal of the extinction coefficient for visible light (K_{vis}), and the percentage of leaves having inclinations to the horizontal greater than 60° are shown in Fig 4.10. There were significant differences between the grasses of contrasting leaf arrangement in their extinction coefficients which led

104 CLIMATIC RESOURCES AND ECONOMIC ACTIVITY

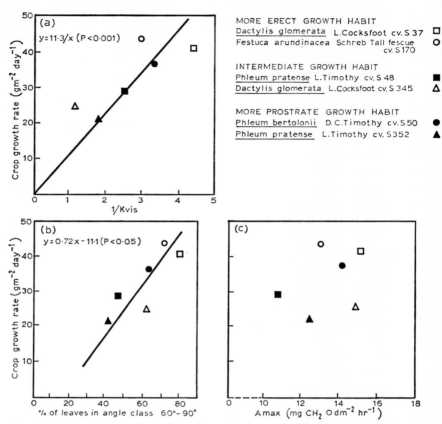

Fig 4.10 Comparison of the effect of species and growth habit in pure grass swards on the relationship of crop growth rate and three variables

to higher crop growth rates. The effect of P_m was small and not significant.

Conclusions

During autumn and early spring, temperature is shown to exert a considerable influence on the rate of leaf expansion, and therefore on crop growth rate. In particular, the artificial raising of the soil temperature in the region of the vegetative shoot apex increased leaf growth and dry-matter production. Although it is not anticipated that it would be economical for the farmer to employ such methods to increase his yield, the plant breeder may be able to use such information in the search for more productive grass varieties. Indeed, Cooper (1964) has shown that considerable variation in the response of leaf growth to temperature is present in various climatic races. Although races which can maintain leaf growth at relatively low temperatures are often susceptible to frost damage, Daday (1963) has shown that winter hardiness and autumn growth in lucerne are controlled by separate genes. Now that we have established the importance of temperature in the meristem region, it should subsequently be possible to select and breed genotypes which will have rates of leaf area expansion at low temperatures comparable with those in our material grown in the heated areas.

The importance of the rate of individual leaf photosynthesis on crop growth is demonstrated by the model, and this character has been shown to have a high heritability (Cooper and Wilson (1970)). Marked differences in photosynthetic rates at low temperatures have been reported—see Treharne and Eagles (1970), Charles-Edwards *et al* (1971); it should therefore be possible to select material which can photosynthesise more effectively during the low temperatures of British winter (Cooper and Breese (1971)). Finally, the model indicates that an increase in L results in a better utilisation of available light energy, particularly when accompanied by a suitable angular arrangement of leaves.

Acknowledgements
We are very grateful for the interest taken in this work by Mr K. M. Cottrell and Mr A. Windram of the Biometrics Department at Hurley and for helpful discussions with them. Part of the work includes results obtained while one of us (J.E.S.) was studying for a PhD at the Welsh Plant Breeding Station, Aberystwyth.

References

BARNARD, C. 'Grass, Grazing Animals, and Man in Historic Perspective', in Barnard, C. (ed), *Grasses & Grassland* (1964), 1–12.

BEEVERS, H. 'Respiration in Plants and Its Regulation', IBP/PP Technical Meeting, *Productivity of Photosynthetic Systems Models and Methods* (Trĕbŏn, 1969), 54–69.

BULL, T. A. 'Expansion of Leaf Area per Plant in Field Bean (*Vicia faba* L) as Related to Daily Maximum Temperature', *J Appl Ecol*, **5** (1968), 61–8.

CANHAM, A. E. 'Electrical Soil Warming', *Electricity in Horticulture* (1964), 51–9.

CHARLES-EDWARDS, D. A., CHARLES-EDWARDS, J. and COOPER, J. P. 'The Influence of Temperature on Photosynthesis and Transpiration in Ten Temperate Grass Varieties Grown in Four Different Environments', *J Exp Bot*, **22** (1971), 650–61.

COOPER, J. P. 'Climatic Variation in Forage Grasses. 1 Leaf Development in Climatic Races of *Lolium* and *Dactylis*', *J Appl Ecol*, **1** (1964), 45–62.

COOPER, J. P. and BREESE, E. L. 'Plant Breeding—Forage Crops and Legumes', in Cooper, J. P. and Wareing, P. F. (eds), *Potential Crop Production* (1971), 295–318.

COOPER, J. P. and TAINTON, N. M. 'Light and Temperature Requirements for the Growth of Tropical and Temperate Grasses', *Herb Abstr*, **38** (1968), 167–76.

COOPER, J. P. and WILSON, D. 'Variation in Photosynthetic Rate in *Lolium*', *Proc 11th Int Grassld Congr, 1970* (Surfers Paradise, 1970), 522–7.

COSTES, C. 'The Energy Requirement of Plants', *Span*, **7** (1964), 15–17.

DADAY, H. 'Frost Resistance and Winter Growth in Lucerne', *Rep Div Pl Ind*, CSIRO, Australia (1963), 15, 23.

HUNT, L. A. 'Ash and Energy Content of Material from Seven Forage Grasses', *Crop Sci*, **6** (1966), 507–9.

KLEINENDORST, A. *Personal Communication* (1972).

KLEINENDORST, A. and BROUWER, R. 'The Effect of Temperature on Two Different Clones of Perennial Ryegrass', *Jaarb Inst Biol Scheik Onderz LandbGewass* (1965), 29–39.

——. 'The Effect of Temperature on the Root Medium and of the Growing Point of the Shoot on Growth, Water Content and Sugar Content of Maize Leaves, *Neth J Agr Sci*, **18** (1970), 140–8.

LANGRIDGE, J. and MCWILLIAM, J. R. 'Heat Response of Higher Plants', in

Rose, A. H. (ed), *Thermobiology* (London and New York, 1967), 231–92.
LEOPOLD, A. C. *Plant Growth and Development* (1964), 369–92.
MITCHELL, K. J. 'Influence of Light and Temperature on the Growth of Ryegrass (*Lolium* spp). 1 Pattern of Vegetative Development', *Physiologia Pl*, **6** (1953), 21–46.
MONTEITH, J. L. 'Light Distribution and Photosynthesis in Field Crops', *Ann Bot*, **29** (1965), 17–37.
OSMAN, A. M. 'Dry Matter Production of a Wheat Crop in Relation to Light Interception and Photosynthetic Capacity of Leaves', *Ann Bot*, **35** (1971), 1017–35.
PATEFIELD, W. M. and AUSTIN, R. B. 'A Model for the Simulation of Growth of *Beta vulgaris* L', *Ann Bot*, **35** (1971), 1227–50.
PEACOCK, J. M. 'Interaction between the Sward and the Environment in the Field', *Ann Rep Grassld Res Inst, 1970* (1971), 55–7.
——. 'Interaction between Sward and the Environment in the Field', *Ann Rep Grassld Res Inst, 1971* (1972), 47–9.
PEACOCK, J. M. and STILES, W. 'Interaction Between a Crop and the Environment in the Field', *Ann Rep Grassld Res Inst, 1968* (1969), 51–3.
RABINOWITCH, E. and GOVINDJEE. 'Solar Energy and Its Utilisation', in *Photosynthesis* (New York, 1969), 38–41.
ROBSON, M. J. 'A Comparison of British and North African Varieties of Tall Fescue *Festuca arundinacea*. 1 Leaf Growth during Winter and the Effect on It of Temperature and Day Length', *J Appl Ecol*, **4** (1967), 475–84.
——. 'Light, Temperature and the Growth of Grasses', *Ann Rep Grassld Res Inst, 1968* (1969), 111–23.
ROY, M. G. and PEACOCK, J. M. 'Seasonal Forecasting of the Spring Growth and Flowering of Grass Crops in the British Isles', in Taylor, J. A. (ed), *Weather Forecasting for Agriculture and Industry* (Newton Abbot, 1972).
RYLE, G. J. A., BROCKINGTON, N. R., POWELL, C. E. and CROSS, B. 'Measurement and Prediction of Organ Growth in a Uniculm Barley', *Ann Bot*, **37** (1973) 233–46.
SHEEHY, J. E. *Studies in Light Interception in Herbage Canopies*, PhD Thesis, UCW (Aberystwyth, 1970).
SHEEHY, J. E. and COOPER, J. P. 'Light Interception, Photosynthetic Activity and Crop Growth Rate in Canopies of Six Temperate Forage Grasses', *J Appl Ecol*, **10** (1973), 239–50.
TREHARNE, K. J. and EAGLES, C. F. 'Effect of Temperature on Photosyn-

thetic Activity of Climatic Races of *Dactylis glomerata* L', *Photosynthetica*, **4** (1970), 107–17.

WATSON, D. J. 'Size, Structure and Activity of the Productive System of Crops', in Cooper, J. P. and Wareing, P. F. (eds), *Potential Crop Production* (1971), 76–88.

WATT, W. R. 'Role of Temperature in the Regulation of Leaf Extension in *Zea mays*', *Nature*, **229** (1971), 46–7.

WENT, F. W. 'The Experimental Control of Plant Growth. Walthan, Mass', *Chronica Botanica* (1957), 343.

WOODFORD, E. K. *Grass and Forage Crops in Grassland Management* (1969), 11.

CHAPTER 5 W. H. HOGG

The Use of Climatic Information in the Classification of Agricultural and Horticultural Land

Introduction

The title of this paper may imply that we know how best to use climatic information in land classification. This is far from the truth and the objective is to state what has so far been done and to make a few tentative suggestions without being able to offer any comprehensive solutions. The paper modifies an earlier report (Hogg (1962)) which was prepared before some of the existing classifications had been devised.

Some restriction must be placed on the type of classification considered. We are not concerned with purely factual surveys on crop distribution, but with the potential for production within the British Isles and the variations in that potential. Three post-war examples may be cited, (i) the Surveys of Potential Horticultural Areas of England and Wales and (ii) the Agricultural Land Classification of England and Wales, both under the direction of the present Lands Arm of the Agricultural Development and Advisory Service; also (iii) the work on Land Use Capability by the Soil Survey.

All of these aim to provide maps of assessments of the quality of land whatever its present use. They are concerned only with the more or less permanent physical factors which go to make up land quality and ignore, for example, transport, nearness to markets, the standard of fixed equipment and the level of management. They are intended to be valid for long periods and, although their main value is for planning the allocation of land among various competing interests, they can be used in advisory work.

Classifications Used in Surveys
Our main interest is with the climatic content of classification but we must first consider the overall systems of grading used in surveys. As an example we shall take that now used in the Agricultural Land Classification, details of which are given in the Explanatory Note (MAFF (1968)).

The land is graded according to the degree to which its physical characteristics impose long-term limitations on its agricultural use. The limitations operate in one or more of four principal ways: they may affect (a) the range of crops which can be grown, (b) the level of yield, (c) the consistency of yield, and (d) the cost of obtaining it. The physical factors taken into account are climate (particularly rainfall, transpiration, temperature and exposure), relief (particularly slope) and soil (particularly wetness, depth, structure, stoniness and available water capacity).

A system of five grades is used:

- Grade I Land with very minor or no physical limitations to agricultural use.
- Grade II Land with some minor limitations which exclude it from Grade I.
- Grade III Land with moderate limitations due to the soil, relief or climate, or some combination of these factors which restrict the choice of crops, timing of cultivation or level of yield.
- Grade IV Land with severe limitations due to adverse soil, relief or climate, or a combination of these.
- Grade V Land with very severe limitations due to adverse soil, relief or climate, or a combination of these.

The Land Use Capability Classification (Bibby and Mackney (1969)) is essentially the same as this but Grade V is expanded to take account of the uses to which such land may be put (pasture, rough grazing, forestry and recreation). As a result this classification is based on seven classes instead of five.

Each of these systems is concerned with limitations and this must be remembered when discussing the climatic factors involved.

The Importance of Climate in Agricultural Classification

So far the climatic factors in classification have not received much attention in Britain. This attitude is not confined to classification but permeates the advice given to farmers. If we accept that the atmosphere comprises about half of the physical environment of a plant and all of the environment of a farm animal, why do farmers ask innumerably more questions concerning the soils than the atmosphere? It seems to be a matter of scale.

Any survey on a world scale emphasises the major influence of climate on crop distribution and may remind us that some earlier climatic classifications made up for their paucity of data by substituting plant distribution. The extensive areas of wheat grown over those parts of the world which were formerly temperate grasslands and the restriction of certain citrus fruits to frost-free areas are clear examples of the relationship between climate and crops.

For surveys covering much smaller areas, the emphasis may be entirely otherwise. While a farmer may grow a wide variety of crops within a small area, he will already have exercised his judgement and excluded many crops from his range of possibilities. No farmer in the British Isles will try to grow a commercial crop of grapefruit and oranges or regard rice as a possible cereal crop. In other words the effects of some of the major climatic controls are automatically limited and the emphasis in decisions concerning crop production is thrown on to the soil. It therefore seems clear that much of the help given by the meteorologist must relate to mesoclimate and this leads to certain practical difficulties.

Climatic Controls on Plant Production

If we accept that crop production is essentially the conversion of solar energy, water and soil nutrients into economic end-products, we may regard solar radiation and precipitation (or other methods of expressing water needs) as primary controls in this process. Also, because they are so closely linked with solar radiation, sunshine, air temperature and soil temperature may be included in this group. A number of other controls can be taken as somewhat less fundamental for crop production in general and these have been discussed fully elsewhere (Hogg (1971)).

Table 5.1 indicates some of these controls, with suggestions as to whether they are largely macroclimatic or mesoclimatic, but such a classification must be somewhat subjective. For example, both solar radiation and frost incidence have their macroclimatic and mesoclimatic aspects, the latter being largely determined by topography. Nevertheless it is worth examining our use of macroclimatic and mesoclimatic information in classifications.

Table 5.1
MAJOR CLIMATIC CONTROLS AFFECTING THE ENVIRONMENTS FOR PLANT PRODUCTION

Primary	Solar energy, precipitation and other methods of expressing water need (macro)
	Sunshine, air and soil temperatures (macro)
Secondary	Advected energy (macro)
	Topographical modifications, eg hill climate (macro, meso)
Tertiary	Weather favourable to plant disease (macro, meso)
	Isolated phenomena causing damage or crop reduction (meso)

Macroclimate

A major difficulty in using macroclimatic data is the assessment of plant requirements. These may be known in very general terms but no

CLASSIFICATION OF AGRICULTURAL AND HORTICULTURAL LAND 113

reasonably precise application is yet possible. For example, the contrast between the predominantly arable areas in the east and grassland to the west of Britain is largely macroclimatic but this is only reflected as guide lines in the present classification. The following table is constructed from the information given in the Explanatory Note for the Agricultural Land Classification (1968) and MAFF Technical Report No 11 (1966).

Table 5.2
MACROCLIMATIC RESTRICTIONS IN CLASSIFICATION (GENERALISED)

Restrictions Lower limits of		
Altitude (ft)	Rainfall (in)	Grading
400	40 (45 in W)	Not above III
600	50	Not above IV
1,000	60	Not above V

Further, there are no rules of thumb and the generalised nature of restrictions must always be remembered. The Land Use Capability Classification adopted similar guide lines and the fact that they are not identical serves to underline the imprecision of the process. In practice, the help at present given by the meteorologist in land classification is very limited. For each of the 1in Ordnance Survey sheets an appreciation is provided which is based on what macroclimatic data are available, very often from stations which are not on the sheet. For some sheets it may be possible to suggest which climatic factors are most important in the grading. Later in this paper some consideration will be given to other parameters which may be useful.

Mesoclimate

If a mapping scale of at least 1in to the mile is used, any land classification concerned with potential production must take note of mesoclimatic variations. The most important of these are likely to be wind

H

and frost (see Table 5.1), but they may also include the penetration inland of sea influences along valleys and possibly the deposition of sea-salt on plants and the effects of industrial pollution. A consideration of local soil types may also lead to some modification of generalised ideas on water requirements.

There is inevitably a large measure of interaction between some of the individual mesoclimatic factors. For instance certain areas along Britain's western seaboards receive the maximum of advected energy and are therefore noted for their earliness, but their location also sharply increases the risk of reduced yields because of exposure to winds. Inland, in horticultural districts, there is often a very delicate balance between the risk of damage to fruit by radiation frost in the valley bottoms and by exposure on the hillsides.

The assessment of mesoclimatic factors in classification presents much difficulty as, almost by definition, the interpolation of macroclimatic data is ruled out, and it is hard to find any substitute for the field assessments used at present. The economic effects of weather and climate are more easily apparent to horticultural growers than to agriculturalists, largely because of the high capital investment per acre and also because the climatic disadvantages may sometimes be overcome. For this reason considerably more help is needed from the meteorologist during classification and in the Survey of Potential Horticultural Areas the meteorologist was a permanent member of the field team which visited a large number of representative sites. He took his place with horticulturists, soil scientists and other advisers in the discussion and assessment of the grading. His contribution rested almost entirely on judgement, which in turn depended on experience and a limited number of detailed surveys of local frost and wind.

At present there are few alternatives. Topographic maps may be used to make crude estimates of the flow of cold air on radiation nights and also to venture general opinions on exposure. These assessments however, are liable to considerable correction in the field.

Another question which arises concerns the cost of overcoming adverse climatic factors such as wind and frost, where this is possible, and the effect on overall grading. Provided the cost is acceptable, it seems reasonable to grade on any remaining disadvantages but large expenditure, as for the provision and maintenance of extensive shelter, should perhaps lead to some downgrading.

Ideas for the Future: Macroclimate

From our definition of crop production given earlier it appears logical that any macroclimatic parameters used in land classification should take account of the energy received by crops and their water needs. Work published in the last few years has some relevance to this problem and a brief account of this follows, although the methods have not so far been used in connection with classification.

Smith (1972) shows the effect of climate and size of farm on the type of farming. On the farming side he deals with the percentage of land under grass and under cereals and also with the number of cattle per hectare. His meteorological parameter is effective transpiration, which is the integration of potential transpiration over the growing period (April to September) at times when the soil moisture deficits within the root range are not great enough to act as a major check to grass growth. It takes into account radiation, temperature, humidity, wind and rainfall and can be regarded as an expression of energy integrated during the time when growth is not limited by a lack of adequate soil moisture. There are very high positive correlations between this climatic factor and (i) the percentage of grass and (ii) cattle per hectare. The correlation with the percentage of cereals is of the same order but negative, and farm size is of greater significance. The general approach could be used in classification but it would require definition as to what are grassland areas and cereal areas. Given these, the average climatic potentials could be calculated.

A second possibility is described by Bibby and Mackney (1969) and is largely based on unpublished work by Dr R. W. Gloyne of the Meteorological Office at Edinburgh. It uses average water balance and temperature during the period April to September, $RP-T$ and $\bar{T}(x)$ where:

R = average rainfall (mm)
PT = average potential transpiration (mm)
$\bar{T}(x)$ = long-term average mean daily maximum temperature

This system is regarded as an inventory of climate rather than a classification and three groups are defined.

Table 5.3
DEFINITION OF CLIMATIC GROUPS

Group	$R-PT$	$\bar{T}(x)$
I	<100mm	>15° C
II	<300mm	>14° C*
III	>300mm	<14° C

* Excluding anything classified as Group I.

Group I is described as having no, or only slight, climatic restrictions to crop growth, Group II refers to areas with a moderately unfavourable climate and Group III to areas with climates ranging from moderately to extremely severe. As the parameters used are based on long-term averages, they are easier to obtain than effective transpiration which requires the computation of water balance sheets. If they are to be used in land classification, however, they must in some way be linked to farming experience. Bibby and Mackney (1969) give climatic inventory data for 106 stations which show that the following very different areas occur in Group I: Penzance, Cranwell, Ushaw and Lossiemouth.

Birse and Dry (1970) have constructed a map of the Assessment of Climatic Conditions in Scotland on a scale of 1:625,000. This is based on accumulated temperature above 5·6° C and the potential water deficit integrated over the period when potential evapotranspiration exceeds rainfall. Eighteen different categories occur in Scotland ranging from warm dry lowland along the coasts of major inlets on the east coast to large areas of cold and very cold wet upland and mountain, from the Grampians northward, and extremely cold wet mountain in the highest parts, especially around Braemar. The map characterises sites where vegetation records have been made and the soils described and sampled. However, it could possibly be used in land classification, although as with other methods, links will need to be forged between the climatic assessments and types of land use.

Ideas for the Future: Mesoclimate

In the British Isles the mesoclimatic aspects of land-use classification are as important as the macroclimatic. Our aim is therefore to accumulate sufficient formal observations or other knowledge to make possible an assessment of local climate anywhere. The general method of achieving this has been to conduct local studies to indicate the variations of the climatic elements which may occur within restricted areas. Its great disadvantage is the considerable time needed to obtain a series of weather samples from a small area; even then, we are often not certain of the general validity of these samples. Clearly, any method of obtaining such information more rapidly will provide more effective help in land classification and no doubt in other spheres.

First in importance is the possibility of undertaking representative investigations. The application of mesoclimatic data would be much extended if the results of investigations could be readily extrapolated to similar topographic situations by studying groups of representative sites selected in relation to land-form. Probably geological, geographical or geomorphological units could be used, for example the effect of dip and scarp slopes on local climates; others may be related to angle of slope and aspect and to slopes of different shape. Some work has already been done on the (theoretical) amounts of radiation these receive but not on the resulting effects on air temperature and soil temperature (see Knoch (1963)). For each representative unit we may also have to take account of generalised soil types and different covers, eg bare soil, arable crops, grass and trees.

Included with this approach are the changes which take place with increasing distance inland from the coast. Gradients of temperature, humidity and sunshine are usually smoothed on the basis of data from two or three stations and the changes may in reality be quite different.

We may also question the validity of the generally accepted lapse-rate of air temperature. The often quoted value of about 3° F/1,000ft (0·6° C/100m) is probably untrue for the lowest few hundred feet (which contains a large proportion of our high quality land) and needs verification for greater heights (see Chapter 2, herein).

Some difficulties would disappear with a denser macroclimatic network, but the existing network could be made more useful in the assessment of local climate. If suitable stations were made the centre of temporary mesoclimatic networks and the data used in the preparation of local climatological maps, interpolation between the macroclimatological stations would become much more realistic.

The second way in which we might increase our knowledge of mesoclimate is by using non-instrumental methods. Must we remain tied to the concept of collecting daily observations over a period of time, or are there any circumstances in which we may obtain a synoptic view of given situations at a point in time to help in the estimates of mesoclimates? One possibility is to use aerial photographs now available in libraries, supplemented if necessary by new ones. If these were taken on mornings with patchy radiation mists and fogs, much could be discovered concerning the distribution of cold air and the susceptibility of individual valleys to radiation frosts. In more general terms we may be able to determine the distribution of donor and recipient areas for cold air (Bush (1945)). Other possibilities include mapping the extent of frost damage, using aerial surveys made a few days or weeks after widespread frost. Perhaps also it would be possible to obtain some hint of exposure in broad bands. Considering how much military information can be deduced from aerial photographs, perhaps a series of photographs at intervals could, by integrating physical and biological factors, enable reasonable deductions to be made on plant environment.

Another suggestion in this field is to make full use of maps when delineating areas of radiation frost, by making some assumptions on the flow of cold air down natural slopes and checking these in the field. If, for example, we can establish that katabatic flow is unlikely unless a certain gradient is reached, we can delineate donor and recipient areas; we can go further and reasonably define recipient areas as subject to frequent radiation frost. In a purely empirical manner this has been done for Somerset using the 1:25,000 scale maps (Hogg (1965)). Although the assumptions have not been tested in the field much advisory work in relation to frost has been with farms near this dividing line. Another possibility is to obtain reliable figures on the varying depth of freezing air during radiation nights and the frequency of the occurrence of the phenomenon. This would require some instrumentation and routine observation, but the results would provide a synoptic outlook on the problem.

Maps could perhaps also be used in defining areas providing large-scale geomorphic shelter. At present there is no means of assessing the shelter effects of a line of hills although there is much information on the shelter provided by smaller topographic features such as shelter belts. Some firm criteria are needed from field work, perhaps combined with wind-tunnel experiments. This would certainly improve our ideas on exposure and shelter as affecting land use.

Some of these ideas may be somewhat far-fetched but unless progress is made along some unorthodox lines it may be that a speaker at the sixty-fifth Aberystwyth symposium will be complaining of the lack of data which can be readily applied in classifying agricultural and horticultural land and also in many other spheres of climatology.

References

BIBBY, J. S. and MACKNEY, D. *Land-Use Capability Classification*, Soil Survey Tech Mon, No 1 (1969).

BIRSE, E. L. and DRY, F. T. *Assessment of Climatic Conditions in Scotland*, 1, based on accumulated temperature and potential water deficit, map and explanatory pamphlet, Soil Survey of Scotland (1970).

BUSH, R. *Frost and the Fruitgrower* (1945).

HOGG, W. H. *The Importance of Climate in Agricultural Land Classification*, Paper No 3 of Technical Report No 8 (Classification of Agricultural Land in Britain), Ministry of Agriculture, Fisheries and Food, Agricultural Land Service (1962).

———. 'Climatic Factors and Choice of Site with Special Reference to Horticulture', in Johnson, C. G. and Smith, L. P. (eds), *The Biological Significance of Climatic Changes in Britain*, Institute of Biology Symp No 14 (1965), 141–55.

———. 'Regional and Local Environments', in Wareing, P. F. and Cooper, J. P. (eds), *Potential Crop Production* (1971), 6–22.

KNOCH, K. 'Die Landesklimaaufnahme Wesen and Methodik', *Ber dtsch Wetterdienstes Nr*, **85** (1963), 1–64.

MINISTRY OF AGRICULTURE, FISHERIES AND FOOD. *Agricultural Land Classification*, Technical Report No 11, Agricultural Land Service (1966).

———. *Agricultural Land Classification Map of England and Wales*, explanatory note, Agricultural Land Service (1968).

SMITH, L. P. 'The Effect of Climate and Size of Farm on the Type of Farming', *Agr Met*, **9** (1972), 217–23.

CHAPTER 6   V. B. PROUDFOOT

The Northern Limit of Agriculture in Western Canada

Introduction

The development of agriculture and settlement in western Canada has been greatly influenced by views of the climate, both at a personal and at a public, governmental, level. As early as 1857 a map showing the supposed northern limits of wheat cultivation was published as a contribution to the assessment of the resources of the area (Dunbar, 1973). Data on which to base such an assessment were extremely scanty, but neither then, nor since, has paucity of reliable information deterred the propagandist.

To assist in the understanding of land productivity and in the rational allocation of resources, objective methods are needed for determining the climatic suitability of a region for various crops. Within a region, climate and daylength limit the crops which may be grown, while climate is mainly responsible for yearly variations in yields. Whether or not a particular crop will be grown within the broad limits set by climate and daylength depends on a variety of cultural factors amongst which profitability is of prime importance to the commercial farmer.

Crop Limits

Let us assume that we can define an isoline north of which a crop cannot grow since a particular limiting climatic condition occurs each year. South of this line we would expect a series of zones within which the crop might grow successively 1 year in 10, 2 years in 10, 3 years in 10 ... until we reached another isoline along which the limiting condition never occurred. If the price obtained by the farmer for his crop were sufficiently high then he might consider it worthwhile to grow the crop near the northern isoline even though he only obtained a crop once or twice in 10 years. Alternatively, if the price were low, he might consider it worthwhile to grow the crop only near the southern isoline where he could depend on obtaining a crop almost every year.

In the real world the situation is more complex than this simple example would suggest. Even when we can define the northern isoline polewards of which the crop cannot grow, rarely can we make unequivocal statements about crop losses south of the line. Instead of having a zone within which the crop will mature, say, 3 years in 10, we are more likely to have a zone in which, during a ten-year period, (a) in 3 years, all the crop matures and produces average or better yields, (b) in 4 years, most of the crop matures but yields are average or a little below, and some of the crop fails completely, and (c) in 3 years, most of the crop fails completely but some matures although yields are poor, and overall yield is very low.

More generally, variations in crop survival and yield in any one year will depend on such local factors as site microclimates, soils, and cultivation practices, such as date and depth of sowing, and fertiliser application. Whether or not a farmer will plant a crop at any particular location depends on his estimate of the risk involved, a risk which can scarcely be formulated in other than rather general terms. To complicate matters further, his estimate of the risk will be based not only on the relatively simple variations of climate, but will be based also on his estimate of prices, and on his expectations. If his expectations are low he may be prepared to accept high risks, for even in the worst circumstances which he envisages, he will still have sufficient to meet his requirements.

Western Canada
The northern limits of contemporary agriculture in western Canada stretch raggedly north-west from the Lake of the Woods, just north of the forty-ninth parallel in south-eastern Manitoba, to Lesser Slave Lake

in north central Alberta at latitude 55·5° N. Farther north-west is an isolated block of continuously settled farmland in the Peace River Region which stretches north beyond Fort Vermilion to almost 59° N. This limit is continually fluctuating—new farmland is being cleared while other land is being abandoned. Small isolated patches of cleared and cultivated land are found north of the limit while larger patches of uncleared land are found south of it.

Along this whole northern limit the major climatic hazard is the short, variable and unreliable frost-free season, although other problems such as difficult terrain, poor drainage, relatively infertile soils or limited access are locally significant (Vanderhill (1962)). The problems of the short, frost-free season were recognised more than a century ago by Dr James Hector, a member of Palliser's Expedition to the Canadian West in 1857–60. He partly attributed the preservation of the rich green pasture in the plains along the North Saskatchewan River to 'the sharp frosts in August and September which arrest the sap before the grasses have fully flowered'—see Hector (1861) and Warkentin (1964, 166). In his introductory remarks to *The Journals, Detailed Reports and Observations* (1863), Palliser himself commented on the paucity of climatic data and noted that severe night frosts occurred frequently in June, rendering the wheat crops very precarious in the already settled Red River and Assiniboine district of southern Manitoba (Warkentin (1964, 181–2)). Palliser continued 'but the climate is well suited to the growth of barley, oats, potatoes, and garden vegetables'. Later he commented on the agricultural capabilities of the Fertile Belt lying between the 'arid' areas of what is now south-east Alberta and the forests to the north:

> The capabilities of this country and its climate, for the success of the cereals, have hardly been sufficiently tested. But I have seen first-rate specimens of barley and oats grown at many of the forts. Wheat has not been so successful but I am hardly prepared to say that this was because of the unfitness of the climate to produce it. I have much reason to believe that the seed has been bad, and the cultivation neglected, and the spots

chosen not of a suitable aspect. I have not only seen excellent wheat, but also Indian corn (which will not succeed in England or Ireland) ripening on Mr Pratt's farm at the Qu'Appelle lakes [in southern Saskatchewan], in 1857. (Palliser, 1863; Warkentin, 1964, 190.)

In this same year, 1857, Lorin Blodget published his great work on the climate of the United States in which he drew attention to the favourable climate of the plains north and west of Lake Superior. Showing that the isotherms are deflected northward on passing Lake Superior westward, he stated that the potential for settlement in this region had been neglected. His comment that the spring opens at nearly the same time along the immense line of plains from St Paul to the Mackenzie River was frequently quoted in later handbooks and guidebooks for intending settlers in the Canadian west (for example, Canada, Dept of Agriculture (1885, 96)). On occasions, as in Hind (1859), this warming effect was attributed to Pacific air crossing the Rockies—'The Prairies enjoy, too, north of the fifty-eighth parallel, the genial, warm and comparatively humid winds from the Pacific, which are felt as far north as the latitude of Fort Simpson [62° N].' (Warkentin (1964, 217).) The long persistence of the belief in the similarity of the climate over this considerable latitudinal range can be explained, not only by the paucity of climatic data, but also by the delimitation on physiographic maps of a 'great central plain' stretching from the forty-ninth parallel to the Mackenzie Delta on the Arctic Ocean (Gage (1925, 53)).

Among the more optimistic writers later in the nineteenth century was John Macoun, a botanist attached first to the western surveys of the Canadian Pacific Railway, then to various Government Surveys, and finally after 1882, permanently to the Canadian Geological Survey. Since Macoun was a botanist, his opinions on the agricultural possibilities of the Canadian West were highly regarded and his views were widely known, not only through his reports to the Railway and Government, but also through his book on *Manitoba and the Great North West* (1882). He was confident that most of the area between Manitoba in the east, and the Rocky Mountains in the west, and as far north as

57° in the Peace River was suitable for agriculture and stock raising. Climatic difficulties were brushed vigorously aside...

> ... if the mean temperature of the two months July and August reaches 60° wheat cultivation will be profitable... in every part of the territory the temperature exceeds this... I believe that wheat culture will yet extend to within a short distance of Hudson's Bay and down the Mackenzie to latitude 65°. Another physical law fixes the greatest yield near the northern limit of successful growth. This was well exemplified by the wheat obtained in 1875 at Lake Athabasca in latitude 58° 42′ N. (Macoun, 1882, 200.)

Such northern wheat became famous when an entry probably from Chipewyan on Lake Athabasca took the Bronze Medal at the Philadelphia Centennial exhibition in 1876 (Macoun (1882, 198)). Macoun was able in his book to marshal substantial evidence that the climatic hazards could indeed be overcome. G. M. Dawson, whom Macoun quotes, noted that wheat, oats, and barley can be grown in the Peace, '... the only point which may admit of question being to what extent the occurrence of late and early frosts may interfere with growth ...'. (Dawson (1879–80), Macoun (1882, 126), Warkentin (1964, 256).) In comparing the climate of the Peace with that of the Edmonton area Dawson noted that '... in both districts the season is none too long for the cultivation of wheat, but if the crop be counted on as a sure one—and experience seems to indicate that it may—the occurrence of early and late frosts may be regarded with comparative indifference'. (Dawson (1879–80), Warkentin (1964, 256).)

Of the arguments used to discount the frost hazard, perhaps none was repeated more often in the propagandist literature of the period than the beneficial effects of forest clearance and cultivation. In the West, as earlier in Ontario, it was claimed that clearance and cultivation reduced the frost hazard. 'The experience of the early settlers in Ontario was similar to that of the early settlers in Manitoba. We never hear now [in 1886] of this [summer frost] as an objection to Ontario.' (Canada, Dept of Agric (1886, 8).) As further observations were made in the northern parts of the West, especially in the Peace River country, increasing emphasis was also placed on the importance at higher latitudes of the long summer day in hastening the ripening of grain crops. Most important of these crops was wheat, the prime commercial crop of the area, and a crop which has remained enormously significant in the whole Prairie economy (Proudfoot (1972)).

The Limits of Wheat Cultivation
The standard variety of wheat planted in the late nineteenth century was Red Fife, a productive wheat of high milling quality (Buller (1919)). Unfortunately, in years with early autumn frosts, Red Fife was often frozen in the fields and the grain either failed to ripen or was badly damaged by frost. In the hope of overcoming this problem Dr William Saunders, the organiser and first Director of the Dominion Experimental Farms, imported a considerable number of varieties of wheat from abroad, and grew them alongside Red Fife at the various experimental farms across the country. He also initiated a programme of improving wheat by breeding, which led finally to the development of the variety Marquis in 1904, by his son Dr Charles E. Saunders, who had become Dominion Cerealist. The male parent of the cross was Red Fife and the female an early ripening Indian wheat known as Hard Red Calcutta. Marquis ripened earlier than Red Fife, giving good yield and high milling quality. Distribution to the public began in 1909 and within ten years this variety had virtually monopolised western Canadian spring wheat production. Another great advantage connected with earliness in Marquis was the diminished risk of loss from Black Stem Rust, because the grain was likely to be ripe by the time rust developed in the late summer. Marquis retained its dominant position until about 1929 when shorter-season or more disease-resistant varieties became available. It was, however, Marquis which allowed the extensive expansion of commercial grain growing into the more northern areas of the West.

Because commercial grain production was essentially commercial

THE NORTHERN LIMIT OF AGRICULTURE IN WESTERN CANADA 127

wheat production, most discussions on northern climatic limitations have focused on the wheat plant and its responses to frost in spring and autumn, the effect of summer temperatures on plant growth, and the effect of changes in daylength on yields. It is not possible here to do more than indicate some of the suggested limits. In 1907 Sir Frederick Stupart, Director of the Dominion Meteorological Service suggested that the isotherm of 57·5° F for at least two summer months probably marked the boundary of dependable agriculture based on wheat production. O. E. Baker (1928, 402, 409, Fig 159) adopted a summer

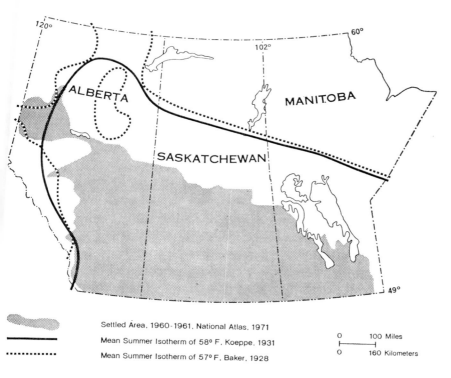

Fig 6.1 Western Canada: mean summer isotherms and extent of settlement

temperature of 57° F as the northern limit of wheat growing, while Koeppe (1931, 115–16) considered the summer isotherm of 58° F to be a safer limit (see Fig 6.1). Whether it is possible to draw such fine distinctions in an area with, even today, a relatively coarse network of meteorological observations is a moot point.

As an alternative to mean summer temperatures, other authors have considered the length of the growing season or the frost-free period. Several writers have argued that the customary map showing the frost-free period is mostly valueless for indicating the northern limits of cultivation, since crops have been grown successfully in areas recorded as having a frost-free season shorter than the period required for maturing crops. Mackintosh, who edited an outstanding series of studies on Canadian Frontiers of Settlement in the 1920s and 1930s, felt that maps showing the period between the average date of seeding and the first killing frost, and the variability in this from year to year, were more significant (Mackintosh (1934, 17, Figs 21 and 22)). Contrariwise, Bennett, as recently as 1959, argued that the 90 day frost-free isoline, using 32° F as a base, provides a relatively accurate northern limit to agriculture, a limit suggested by Reed in 1916, and noted in such standard agronomy texts as that by Leonard and Martin in 1963 (see Fig 6.2).

The length or the warmth of the growing season as expressed in degree days above 42° F, has been considered by a number of writers as providing a significant limit to wheat production. Rodwell Jones' isoline delimiting the 110 days growing season (1928) is broadly similar to the 2,000 degree-days isoline produced by Chapman and Brown (1966) in their study of the *Climates of Canada for Agriculture*. The latter point out that no account is taken in the degree-day concept of daylength. However, already as early as 1912, in a brilliant discussion of the climatic limit for wheat production in Canada, Unstead had suggested combining lengths of daylight with accumulated temperatures to arrive at a limiting isoline. He argued that a meaningful index of the effect of daylength could be obtained by adding mean temperature and mean darkness since higher accumulated temperatures were required at

THE NORTHERN LIMIT OF AGRICULTURE IN WESTERN CANADA

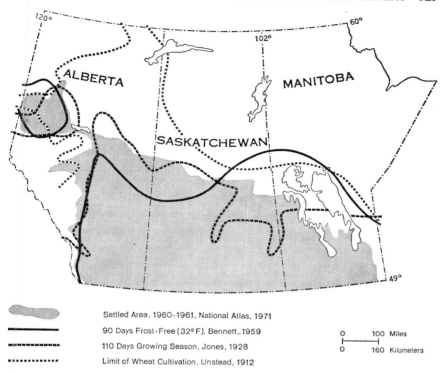

Fig 6.2 Western Canada: mean climatic limits and extent of settlement

locations where the mean temperature was higher and the duration of daylight shorter. The accumulated temperatures were obtained by summing those days after the mean temperature rose above 5° C in spring until it dropped below 10° C in the autumn.

Amongst more recent interpretations of the climatic limits for agriculture are those by Williams (1969) who has rejected a photothermal unit approach in favour of using estimated temperature normals and a biometeorological time scale to prepare a wheat zonation map for the Canadian West. Regressions of temperature normals at nearly 1,200
I

locations were obtained based on latitude, longitude and elevation, ultimately derived from a limited number of observations across the region. In Alberta north of Edmonton, areas within which wheat would be expected to mature are confined to the river valleys, while the northernmost suitable area would be as far as 63° N in the Mackenzie Valley. No limits are shown for northern Saskatchewan and Manitoba.

Meso- and Microclimatic Variations

Absent from most of the agro-climatic interpretations cited above are (a) the problems induced by meso- and microclimatic variations, especially those dependent on soil and local relief differences, and (b) the problems of the economic viability of wheat, or other grain, production in such northerly locations. Two fundamental assumptions underlying the cited discussions are that means or climatic normals are agriculturally significant, and that the meteorological stations are representative. As Mackintosh (1934, 17–18) commented:

> The agricultural settler, however, does not live by averages, annual or seasonal, but from day to day and from season to season. He deals with climate as a matter of probability . . . The variability of climatic conditions is more important than their average occurrence. It is possible for crops to be frozen three years out of five in an area which has an *average* frost-free period longer than the required growing season.

Moreover, those areas which produce the highest yields on average, do not necessarily do so at the least average cost of production, as Craddock (1970, Figs 3.1–3.4) has recently demonstrated for grain production in western Canada. Areas with high average yields may indeed be areas of below-average incomes (Canada, Dept of Forestry (1964, Map 3)). Reminding ourselves of the problems that were raised at the very beginning of this paper, it is not too much to stress that climatic normals must be regarded with considerable care when they are used for purposes of agricultural interpretation.

The second major assumption that meteorological stations are representative is one that must also be challenged. As Longley (1967), Longley and Louis-Byne (1967) and Hayter (1972) have pointed out, representativeness has both spatial and temporal dimensions. Short-term climatic fluctuations of the order of a decade in length are well recognised phenomena, although their effects on station norms are rarely discussed. For example, if a 30 year record for a station included 2 decades with long frost-free periods, and one decade with short frost-free periods, the long-term mean could be substantially different from a mean derived from two decades with short frost-free periods, and one decade with long frost-free periods. The farmer is more concerned with the actual frost-free period over a few years than with the mean over the long-term. This problem is clearly of greatest concern where the frost-free period is marginal for production, and Longley (1967) has been able to show that in Alberta the mean frost-free period for the years 1951–64 differed from the mean for the years prior to 1951. There were differences between adjacent river basins within the plains area of the province, the greatest difference occurring where the frost-free period had increased after 1951, by 3 or 4 weeks, for example, from 80 days to 113, and from 100 to 121.

Within a relatively small area of north-east Alberta, Hayter (1970, 1972) has been able to demonstrate that the spatial variation of temperatures is such that no one particular station can be regarded as representative of the area. The records of the 2 most favourable stations are appropriate only for extremely small localities within the area. On the other hand the 3 stations with the least favourable frost-free periods did not give the absolutely shortest periods that are known to occur in the area. On the basis of this work it is reasonable to conclude that several localities within the area record frost every month while extensive tracts have only 70–105 days frost-free. Between adjacent meteorological stations with an average variation of 5° F in mean monthly minimum temperatures there was a difference in the frost-free period of 57 days. Many of these differences are explicable in terms of topographic variation rather than absolute elevation and exposure.

Conclusion

Detailed local studies emphasise the difficulties of drawing meaningful isolines to define the climatic limits of crop production. Further difficulties arise when other environmental constraints, such as soils, are considered. Perhaps, however, the greatest problem lies not in the drafting of such limits by the impartial observer, but in the use of such limits by policy makers and propagandists who fail to appreciate, or disregard, the assumptions and uncertainties that underlie them.

References

BAKER, O. E. 'Agricultural Regions of North America, Part VI—The Spring Wheat Region', *Econ Geog*, **4** (1928), 399–433.

BENNETT, M. K. 'The Isoline of Ninety Frost-Free Days in Canada', *Econ Geog*, **35** (1959), 41–50.

BLODGET, L. *Climatology of the United States* . . . (Philadelphia, 1857).

BULLER, A. H. R. *Essays on Wheat* . . . (New York, 1919).

CANADA, DEPT OF AGRICULTURE. *A Guide Book Containing Information for Intending Settlers, with Illustrations* (Ottawa, 1885; 6th edn).

——. *Canada: Its History, Productions and Natural Resources* (Ottawa, 1886).

CANADA, DEPT OF FORESTRY. *Economic and Social Disadvantage in Canada* (Ottawa, 1964).

CHAPMAN, L. J. and BROWN, D. M. *The Climates of Canada for Agriculture*, Canada Land Inventory, Report No 3, Department of Forestry and Rural Development (Ottawa, 1966).

CRADDOCK, W. J. *Interregional Competition in Canadian Cereal Production*, Economic Council of Canada (Ottawa, 1970).

DAWSON, G. M. 'Peace River', *Geological and Natural History Survey of Canada, Report of Progress 1879–80* (Montreal, 1879–80), 46B–79B.

DUNBAR, G. S. 'Isotherms and Politics: Perception of the Northwest in the 1850's', in Rasporich, A. W. and Klassen, H. C. (eds), *Prairie Perspectives 2* (Toronto, 1973), 80–101.

GAGE, W. J., & Co. *Public School Geography* (Toronto, 1925).

HAYTER, R. *The Frost Hazard for Agriculture in Northeast Alberta*, unpublished MA Thesis in Geography, Univ of Alberta (Edmonton, 1970).

——. *Spatial Variations in Minimum Temperatures in Northeast-Central Alberta* (Environment Canada, 1972).

HECTOR, J. 'Physical Features of the Central Part of British North America . . .', *Edinburgh New Philosophical J*, NS, **14** (1861), No II, 216–22.

HIND, H. Y. *Report on the Assiniboine and Saskatchewan Exploring Expedition to the Legislative Assembly, Canada* (Toronto, 1859).

JONES, Ll. R. 'Some Physical Controls in the Economic Development of the Prairie Provinces', *Geography*, **14** (1928), 284–302.

KOEPPE, C. E. *The Canadian Climate* (Bloomington, Illinois, 1931).

LONGLEY, R. W. 'The Frost-Free Period in Alberta', *Can J Plant Sci*, **47** (1967), 239–49.

LONGLEY, R. W. and LOUIS-BYNE, M. *Frost Hollows in West Central Alberta*, Meteorological Branch Tech Circ (Ottawa, 1967).

MACKINTOSH, W. A. *Prairie Settlement—the Geographical Setting* (Toronto, 1934).

MACOUN, J. *Manitoba and the Great North-West* (Guelph, Ontario, 1882).

MARTIN, J. H. and LEONARD, W. H. *Principles of Field Crop Production* (New York, 1967; 2nd edn).

PALLISER, J. *The Journals, Detailed Reports, and Observations Relative to the Exploration by Captain Palliser . . . Presented to Both Houses of Parliament by Command of Her Majesty, 19th May 1863* (1863).

PROUDFOOT, B. 'Agriculture', in Smith, P. J. (ed), *The Prairies* (Toronto, 1972), 51–64.

REED, W. G. 'The Probable Growing Season', *Monthly Weather Review*, **44** (1916), 509.

STUPART, R. F. 'Evidence . . . and Short Report Regarding the Climate of Canada', in *Evidence Heard before a Select Committee of the Senate of Canada, 1906–7, . . .* published as *Canada's Fertile Northland* (Ottawa, 1907), 131–9.

UNSTEAD, J. F. 'The Climatic Limits of Wheat Cultivation, with Special Reference to North America', *Geog J*, **39** (1912), 347–66, 421–46.

VANDERHILL, B. G. 'Observations in the Pioneer Fringe of Western Canada', *J of Geog*, **61** (1962), 13–20.

WARKENTIN, J. *The Western Interior of Canada* (Toronto, 1964).

WILLIAMS, G. D. V. 'Applying Estimated Temperature Normals to the Zonation of the Canadian Great Plains for Wheat', *Canadian J Soil Sci*, **49** (1969), 263–76.

CHAPTER 7   J. C. RODDA

Water Resources in the United Kingdom: a Hydrological Appraisal

Introduction

Water is the natural resource which is most universally and constantly in demand. This demand is increasing all over the world. In regions where water is scarce, available resources have always been carefully husbanded to avoid the disastrous alternative. Only in the recent years has there been any doubt that British resources are not able to meet all the demands made upon them. Consumption of water is increasing and it is expected that by AD 2000 the UK demand will be more than twice what it is today. This rising consumption must be set against (a) the decreasing number of inland sites available for reservoir construction, (b) pressures for use of resources for amenity purposes, (c) an increasing proportion of treated effluent in river flows and (d) a rising level of pollution.

However, water resources need not be thought of solely in relation to public water supply, but in the rather wider context of any large-scale activity or project involving water and its use in some way. These range from transport to flood control and hydro-electric production to recreational uses of water such as fishing (Table 7.1). In terms of the capital invested in water resources projects in Britain, some of these items involve only small sums but others represent a considerable outlay. For example, during the financial year 1971/2 capital expenditure on water supply, sewerage and sewage disposal in England and Wales was about £680 million (DOE (1971)). Similarly, the expenditure on

Table 7.1
WATER RESOURCES PROJECTS
(after Dixon)

No	Purpose	Works involved
1	Public water supply	Dams, reservoirs, wells, conduits, pumping stations, treatment works, desalting plants, distribution systems.
2	Irrigation	Dams, reservoirs, wells, canals, pumping stations, desilting works, distribution systems.
3	Flood control	Dams, regulating reservoirs, levées, floodwalls, channel improvements, by-pass channels, pumping stations, zoning, flood forecasting.
4	Drainage	Ditches, tile drains, levées, pumping stations, soil treatment, culverts.
5	Hydroelectricity	Dams, reservoirs, penstocks, power plants, tunnels and shafts.
6	Navigation	Dams, reservoirs, canals, locks.
7	Pollution and safety control	Regulating reservoirs, treatment facilities, barriers, groundwater recharge facilities.
8	River and basin conservation	Soil conservation practices, headwater control structures, fish ladders and hatcheries, land management practices.
9	Recreation	Reservoirs and canals with access for fishing, boating and scenic areas.

road construction and maintenance was considerable, being £630 million for 1969 (Road Research (1969)); part of this expenditure, possibly 3–4 per cent, was on hydraulic structures such as bridges and culverts. In fact, in motorway construction between 7 and 10 per cent of the cost, about £1 million per mile, is devoted to hydrological structures. On the other hand few, if any, projects have been undertaken in this country solely for the recreational facilities they provide, while the expenditure on transport by British Waterways is just over £3 million a year (British Waterways Board (1971)), which is not a particularly large sum in this context.

The Hydrological Network

Of prime importance to the design of any water-resource project is the information collected through the country-wide hydrological network, and employed as a basis for that design. This network consists of not only the gauging stations on rivers and streams where discharge is measured continuously, but also the rainfall stations, the climatological

Table 7.2
HYDROLOGICAL NETWORK CLASSIFICATION

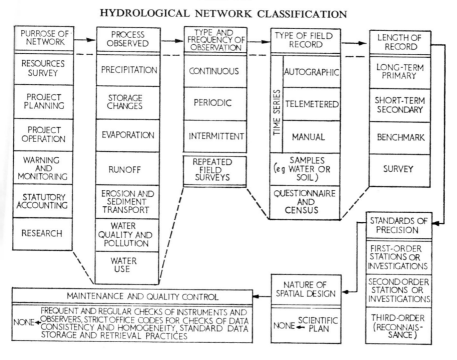

Table 7.3
INFORMATION ON ASPECTS OF NATIONAL HYDROLOGICAL NETWORKS

	Country	Area ($km^2 \times 10^3$)	Per 1,000 km^2 No of rain gauges	Per 1,000 km^2 No of stream gauges		Country	Area ($km^2 \times 10^3$)	Per 1,000 km^2 No of rain gauges	Per 1,000 km^2 No of stream gauges
1	Albania	29	6·7	1·0	31	Korea	66	21·3	2·2
2	Argentina	2,776	1·2	0·3	32	Liberia	111	0·2	1·3
3	Australia	4,928	1·3	0·3	33	Malawi	118	1·8	1·1
4	Austria	83	9·2	3·0	34	Mauritania	1,085	0·03	—
5	Belgium	31	1·1	1·5	35	Mauritius	2	123·9	8·6
6	Brazil	8,850	0·6	0·1	36	Morocco	500	1·2	0·01
7	Bulgaria	111	6·9	2·7	37	Netherlands	37	8·7	6·4
8	Canada	9,210	0·2	0·1	38	New Zealand	268	5·0	1·7
9	Central African Rep	660	0·2	0·007	39	Norway	324	2·2	1·9
10	Chad	1,284	0·2	0·005	40	Pakistan (E & W)	947	1·1	1·2
11	Congo (DR)	2,344	0·2	—	41	Peru	1,285	0·7	0·02
12	Costa Rica	50	2·8	0·6	42	Rwanda	26	0·3	0·3
13	Cyprus	9	1·5	7·6	43	Senegal	201	0·3	0·02
14	Czechoslovakia	128	9·7	4·1	44	Spain	495	6·3	0·9
15	Denmark	43	10·5	1·6	45	South Africa	1,210	1·7	0·6
16	Ethiopia	907	0·1	0·006	46	Sweden	450	2·2	0·9
17	Finland	337	1·7	0·6	47	Switzerland	41	11·2	4·4
18	France	550	7·8	1·4	48	Tanzania	833	1·1	1·6
19	Gambia	10	0·4	—	49	Thailand	514	2·0	1·1
20	Germany (FR)	248	11·9	7·1	50	Tunisia	167	3·1	0·6
21	Ghana	238	1·9	0·1	51	Turkey	776	1·7	1·2
22	Guatemala	131	3·2	0·3	52	Uganda	236	1·8	0·3
23	Guinea	255	0·2	—	53	UAR	1,002	0·09	0·07
24	Honduras	115	0·7	0·4	54	United Kingdom	241	26·8	1·9
25	Hungary	93	4·2	0·6	55	USA	9,300	1·1	0·8
26	Ireland	70	11·9	4·1	56	USSR	22,402	0·8	0·5
27	Israel	21	32·4	5·0	57	Venezuela	912	0·4	0·3
28	Italy	301	6·9	1·7	58	Yugoslavia	256	10·5	2·8
29	Japan	370	16·2	5·4	59	Zambia	752	0·8	0·2
30	Kenya	583	2·56	0·7					

Sources: 'Report of Working Group on Network Design', Commission for Hydrometeorology (Warsaw, 1964); 'Major Deficiencies in Hydrological Data', ECA/WMO (1966); Abstracts of Papers, Water for Peace Conference (Washington, 1967); United Nations, *Statistical Yearbook* (1966).

stations and the wells and boreholes where ground-water levels are logged. In addition, water quality is monitored at a number of points in terms of the dissolved solids present and at some for the suspended load, the biological characteristics and the radioactive constituents. It might be argued that if a network is accepted as being any system for the acquisition of hydrological data, then the term network should not be limited to these station-type time-series observations. Surveys should be included such as the recent River Pollution Survey (DOE and WO (1970)) and even maps, for example soil maps and solid geology maps showing features of hydrological importance. In this sense the country-wide hydrological network is much more comprehensive than may be envisaged at first; in fact it might involve nearly all the elements suggested in this classification of networks. On the other hand it is unusual to find that even part of the network within a single river basin can be identified with just one purpose. Even on this scale the network is dual or multipurpose and its growth has been haphazard rather than scientifically designed.

The rainfall network in Britain is probably the best in the world in terms of numbers of gauges and length of records. This is demonstrated by the 1965 totals in Table 7.3 which shows that the United Kingdom network ranks high in comparison with those of other countries. This position is due largely to the efforts of Symons, Mill and Glasspoole; since 1862 not only has the number of gauges risen (Fig 7.1) but a remarkable degree of uniformity has been achieved in terms of gauge type, site and observer practice. On the other hand neither the number of river gauges in 1965 nor the length of records (Table 7.4) seem to be adequate for as developed a country. There is some doubt, however, whether this is the correct number of gauging stations in operation because the Institute of Hydrology's Floods Team searches have revealed the existence of more than 1,100 river-gauging stations in the United Kingdom, not all of them being in operation at the present. Records from a large proportion of these stations have not been submitted to the Water Resources Board for publication and so they have

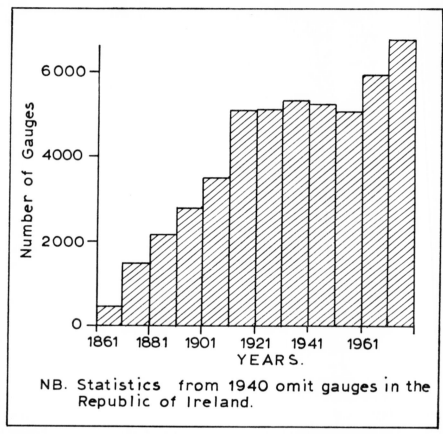

Fig 7.1 Numbers of rain gauges in the United Kingdom, 1861–1971

not been included in the total shown in Table 7.4. Another component of the network is the 641 climatological, agrometeorological and synoptic stations that provide records from which evaporation can be computed. Many of these stations are equipped with rain recorders and, including the other rain-recorder sites that exist, there are approxi-

Table 7.4
UNITED KINGDOM: RIVER GAUGING NETWORK CHARACTERISTICS

Number of stations published in 'Surface Water Year Book'		Lengths of record of stations held in Institute of Hydrology	
No of river gauges	Year	No of stations	Length of record (years)
28	1935	216	0·5–3
52	1936–	97	3–5
	1945	110	7
81	1946–	179	10
	1952	138	15
102	1953	77	20
116	1954	25	25
128	1955	26	30
147	1956	34	35
174	1957	5	40
188	1958	3	45
205	1959	1	50
238	1960	1	55
271	1961	–	60
311	1962	3	65
364	1963	1	70
434	1964	–	75
450	1965		

mately 600 stations in the United Kingdom producing continuous records of rainfall. Ground-water levels are recorded systematically in over 1,000 observations wells in England and Wales and another 200 or more in Scotland (Fig 7.2). The majority of these wells are confined to the Chalk areas but some occur in the Triassic Sandstones and some in other aquifers such as the Jurassic Limestones and Magnesium Limestone.

The different types of information collected by the hydrological network are assembled separately by several organisations and published

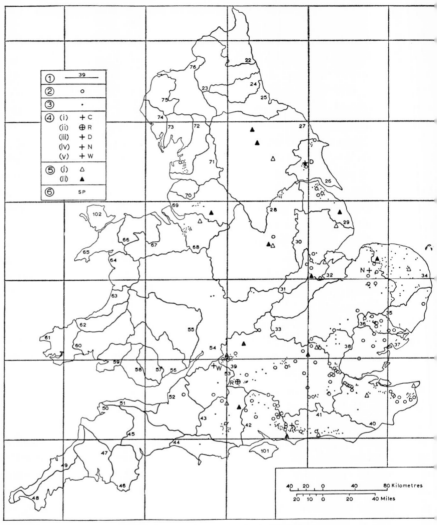

Fig 7.2 The ground-water network in the United Kingdom
(1) Hydrometric area and boundary numbers
(2) Wells recorded automatically
(3) Wells measured manually
(4) Wells for which ground-water hydrographs have geen constructed
(5) Certain Meteorological Office rainfall and weather stations
(6) The 100km square of the national grid

in a number of forms. Rainfall records are collected by the Meteorological Office from the individuals, the river authorities and water boards, the government departments and other bodies that operate rain gauges, and these records have been published annually in *British Rainfall*. Records of river flow are gathered by the river authorities and the river purification boards in Scotland and published annually by the Water Resources Board and the Scottish Development Department in the *Surface Water Year Book*. The *Ground Water Year Book* deals with ground-water fluctuations and is published by the Water Resources Board, whilst the national archive of well records is maintained by the Institute of Geological Sciences. Unfortunately, there is a considerable delay in publishing these year books and *British Rainfall*.

Evaporation is also reported in *British Rainfall*, particularly the records from the rudimentary network of potential evaporation gauges organised through the Nature Conservancy. Potential evaporation is also determined, using the Penman (1948) method for the assessment of irrigation need, and average values have been published in 'The Calculation of Irrigation Need' (MAFF (1954)) and 'Potential Transpiration' (MAFF (1967)). The Penman method is also employed on a routine basis to calculate the state of soil moisture, and maps of the whole country are released at fortnightly intervals by the Meteorological Office, showing the distribution of the soil moisture deficits. One of the other important maps is that of average annual rainfall—the latest published being for the period 1916–50.

A rough estimate has been made (Table 7.5) of the cost of installing and maintaining the UK national network at 1971 prices, water quality monitoring costs being excluded. Most of the investment is in river gauging structures, but the total sum involved seems small by comparison with the ongoing expenditure on water resource projects.

Network Inadequacies

Some of the deficiencies of the UK network have already been noted: they arise from three main sources. These are: first, the difficulties of

Table 7.5
ESTIMATE OF COSTS OF UNITED KINGDOM HYDROLOGICAL NETWORK

(in £)

A CAPITAL (Assuming installation at current prices)

1 Rainfall	Rain gauges and recorders	140,000	
2 Evaporation	Climatological stations	250,000	
		———	390,000
		390,000	
3 River flows	River gauging stations	8,000,000	8,000,000
4 Ground water	Wells and boreholes	550,000	550,000
			8,940,000

B RECURRENT

1 Rainfall	Rain gauge observations	190,000	
2 Evaporation	Climate observations	175,000	
	Rainfall and evaporation processing and publication	400,000	
		765,000	765,000
3 River flows	Maintenance collection analysis and publication	450,000	450,000
4 Ground water	Maintenance collection analysis and publication	75,000	75,000
			1,290,000

measuring the variables involved; second, the low density and poor distribution of stations; and third, the fact that a large part of the records refer to short time-periods. Consider the requirements for flood control purposes as an example. One of the essential features of a flood control project is a system of telemetering rain gauges so that warnings of intense rain can be transmitted to a central control. This allows remedial action to be taken based upon a forecast of flood volumes and levels. Apart from the River Dee basin, where there is a combined system of telemetering rain gauges, water-level recorders and weather stations on line to a computer—and a weather radar in addition—there are few other basins where a similar system operates; yet for the short rivers of Britain such a system would seem essential to establish a sound system of flood forecasting. The fact that the ground-water network is limited to two or three aquifers has already been noted. Moreover, only a small number of these wells and boreholes are equipped with automatic water-level recorders, so that the range of detail required for a number of purposes is not currently available. The remainder of the subsurface network is even more scanty. There are few sites where infiltration is assessed. Soil moisture is continuously monitored at a small number of sites but there has been no attempt to collate this information and publish it. If this information were readily available, it could be employed for directly determining the irrigation-need and for providing additional information on which to base flood forecasts.

One point that has not been discussed is the quality of the data. Obviously a lengthy record is of little use if it is full of errors, but fortunately much of the UK network is regularly inspected and is subjected to numerous checks. Observer practice and instrument installation and maintenance have been standardised, while quality control procedures are operated on most of the records. Obviously, errors can still occur, due to weaknesses of method or observation. For example, records of flood discharges are incomplete at many sites because rating curves have not yet been constructed for large enough flows, and structures can be bypassed or over-topped. Similarly, rainfall records

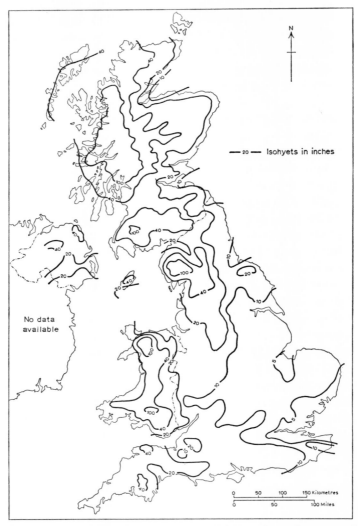

Fig 7.3 United Kingdom: 'residual rainfall', estimated at ground level (minus actual evaporation)

are subject to a bias that varies across the country of between 3 and 20 per cent, because the standard gauge records less than the amount of rain actually reaching the ground on an annual basis (Rodda (1970)). Hence the average annual rainfall map needs appreciable amendment to be realistic hydrologically.

Some Hydrological Studies of Water Resources
(a) *Water Supply*

About 99 per cent of the population of Britain has a piped water supply, by far the largest part being served by public mains. The major areas of demand are obviously the large urban centres, particularly those of the 'axial belt'. Those centres near the 'highland fringe', such as Manchester and Sheffield have relatively easy access to productive, highland, surface-water sources situated where rainfall amounts are large and evaporation rates low. On the other hand, London and the other cities of the south-east are located where the average annual difference between rainfall and actual evaporation is smaller, in other words where run-off amounts per unit area are the lowest for the whole country (Fig 7.3). From the supply point of view, this distribution of demand is one that causes considerable problems. Already abstractions are nearing the quantity authorised to be abstracted under the statutory licensing scheme (Water Resources Board (1972)) and in some areas, local resources will have been fully exploited shortly (Fig 7.4). Many of the sources of supply in the south and east are based directly on groundwater abstraction and even dry-weather flows at surface-water sources consist of discharge of ground water. To maintain these sources, adequate recharge of aquifers is necessary—recharge which takes place mainly during the winter when the excess of rainfall over evaporation reaches a maximum. But obviously there are considerable year-to-year variations of rainfall, and to identify the magnitude of these variations, a probability study was made for some 60 rainfall stations in the south-eastern quarter of the country, each having 50 years of records. Totals were calculated for the five winter months when most recharge occurs,

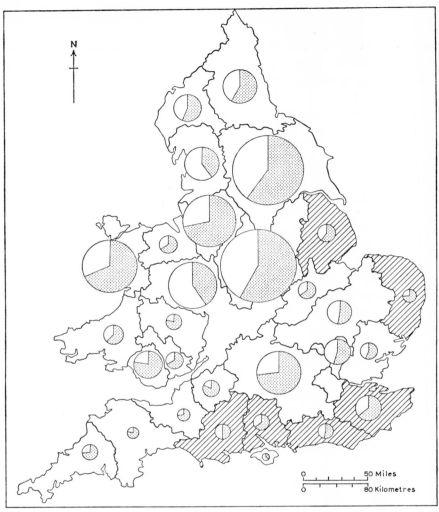

Fig 7.4 England and Wales: abstractions of water; actual compared with authorised. The radius of the circle indicates the amount authorised for abstraction in $M^3 \times 10^5$. *Note:* the shaded portion of the circle is the proportion of actual abstractions. The areas shaded are where ground-water forms the major source of supply

Source: WRB Annual Report for year ending 30 September 1971

namely, November to March, and following a test which showed that these five-monthly totals were sufficiently normal for the purpose of this analysis, rainfall/probability relations were determined for each station. Then a series of maps was constructed to show the distribution of rainfall amounts for particular return-periods (eg Fig 7.5). This map shows how low the amounts of rainfall are likely to be in the south-east, amounts that would induce little percolation and, consequently, very little recharge of the Chalk and the other aquifers. Again, future water demand is forecast to rise very rapidly in south-eastern England (Sharp (1967)), so that by AD 2000 the effects of a 10 year drought could be much worse than a 25 year or 50 year drought occurring now, if new sources of supply are not able to keep up with demand.

(b) *Floods*

During most years the cost of repairing flood damage in Britain amounts to several million pounds. Heavy rain is the usual cause of floods, but rapid snow-melt has caused country-wide flooding in the past. Somerset and Dorset seem particularly prone to heavy falls of rain and questions arise about the return periods of these falls of 225mm (9in) or more and whether it is realistic to ascribe return periods to them. There is also the question of whether there is a limit to the amount of rain that can occur within a given time period at a particular location.

Various attempts have been made to answer this last question, such as by the storm transposition and the probable maximum precipitation (PMP) approaches. The PMP approach, which relies on a physical maximisation of the controlling factors, is well documented and has been applied in various parts of the world. It is not without criticism, however, the whole concept being considered one of expediency by Yevjevich (1968), who suggests that the PMP could always be raised simply by increasing the thickness of the air mass by 1 metre or by making the vertical rise of air a little larger.

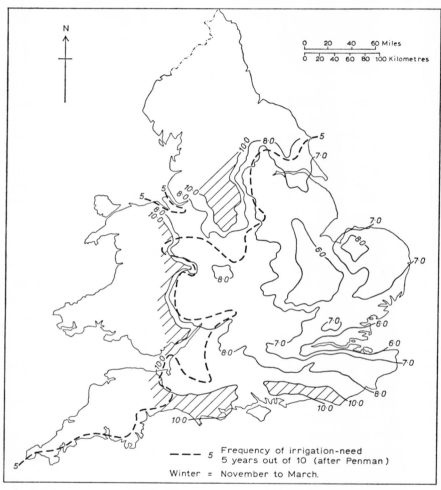

Fig 7.5 England and Wales: total winter rainfall; 100 year return period

An alternative approach was devised by Hershfield (1961) using past records and employing Ven Te Chow's (1951) general frequency formula:

$$X_T = \bar{X}_N + KS_N \qquad (1)$$

where:

X_T = the rainfall for a return period T years when a particular extreme value distribution is used

$\bar{X}_N$ and S_N = the mean and standard deviation for a series of N annual maxima

K = the standardised variate

Following Hershfield's example, records from a frequency study of daily extreme falls for 121 UK rainfall stations (Rodda (1967)) were employed to determine the maximum value for K which would be used in the maximisation. Several K values in excess of 15 were found, but the greatest was for Martinstown, Dorset, where a value of 26·7 was attained. This high figure must be ascribed to the difference between the largest fall and the remainder of the sample, the largest being 279mm (11in), almost five times greater than the second highest daily fall. Using $K = 27$, the PMP was estimated for the other 120 stations using the mean and standard deviation appropriate to each. The resulting map (Fig 7.6) displays a contrast between the drier south and east and the wetter north and west, viz, less than 400mm (15·5in) as against over 750mm (30in) for parts of Snowdonia. Just how realistic these values are is very difficult to judge, but, for techniques of design that demand PMP estimates, they may be sufficiently precise.

Like the risk of drought, the flood risk is likely to grow in the future as flood-plain occupance increases and land-use changes continue. Urbanisation is the land-use change that is considered by most hydrologists to induce rather than attenuate floods, but there is little quantitative evidence to support this hypothesis. The Institute's work at Milton Keynes is designed to show whether such changes do occur. On the other hand there is already some evidence from the Plynlimon

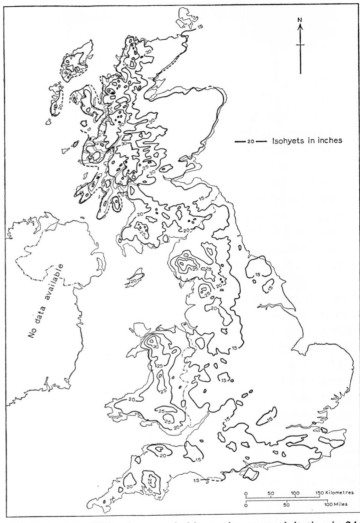

Fig 7.6 United Kingdom: probable maximum precipitation in 24 hours

catchments that mature forest as opposed to sheep pasture delays and reduces flood peaks (Fig 7.7).

The Future
One way of solving many of Britain's water supply problems would be to increase considerably the cost of water, but this has obvious practical difficulties with the present method of charging. Desalination on a large scale and inducing precipitation are possible alternatives but river-regulating reservoirs and estuarial barrages, together with greater re-use of water, artificial recharge and ground-water pumping to stimulate low flows, are the types of scheme that are now being undertaken or are proposed (Water Resources Board (1970a, 1970b)). These schemes and others, including conventional reservoirs, will have to overcome the considerable deficit (Fig 7.8) that is anticipated for the year 2001 (Rydz (1971)) and they will also have to cope with increased demands farther into the future. Improved long-range weather forecasts and the application of hydrological forecasting on an operational scale will assist in mitigating the effects of both floods and droughts as will the improvements to the hydrological network that are continuing. Efficient management of this network may be more difficult in the future without the Water Resources Board or a comparable organisation to provide a national focus, and with ten new regional water authorities concerned with every aspect of water from source to sewer. Langbein (1964) makes a very good case for *not* combining responsibility for the national hydrological network with other functions:

> The collection of basic data must be in an organisation separate from administration or operation. To combine basic data collection with administrative functions usually leads to one or more of the following difficulties:
> (a) Neglect of publication of basic data, and their eventual loss.
> (b) Neglect of collection of data during critical times, as during floods, when personnel may be shunted to other work.
> (c) Uncertainty whether the records are biased. Citizens, or groups,

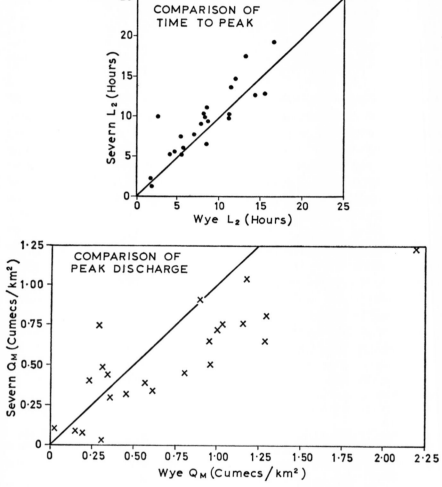

Fig 7.7 Hydrograph characteristics for the Plynlimon catchments

public or private, should have ready access to data from independent sources.
(d) And, most important, record collection emphasizes the main function of administrative agency. For example, an organization concerned mainly with irrigation would give little attention to collection of data needed for design of highway bridges and culverts.

The future may also hold more problems for the management of hydrological research, but it is very evident that such research is most necessary. Water can be termed a natural resource, a climatic resource, or an environmental resource, but its abundance or insufficiency and its spatial and temporal distribution need long-term basic research. This is hydrology. Development of sources of supply, construction of flood prevention schemes and the building of hydro-electric works are technological problems. These problems are civil engineering. They require both *ad hoc* investigations and applied research.

As Langbein (1964) says, there may be many agencies with an interest in hydrological research, but there should be one with basic long-term research as a primary task. This agency can ensure that the research programmes which are carried out, are designed to fill the gaps and look ahead to future needs. It must also have independent authority to publish its findings, whether or not they agree with those of other governmental authorities.

References

BRITISH WATERWAYS BOARD. *Annual Report and Accounts, 1970* (HMSO, 1971).
CHOW, VEN TE. 'A General Formula for Hydrologic Frequency Analysis', *Trans American Geophysical Union*, **32** (1951), 231–7.
DEPARTMENT OF THE ENVIRONMENT AND THE WELSH OFFICE. *Report on the River Pollution Survey of England and Wales*, Vol 1 (HMSO, 1970).
DEPARTMENT OF THE ENVIRONMENT. *Reorganisation of Water and Sewage Services: Government Proposals and Arrangements for Consultation*, Circular 92/71 (1971a).
——. *Report of a River Pollution Survey of England and Wales, 1970* (1971b), 1.

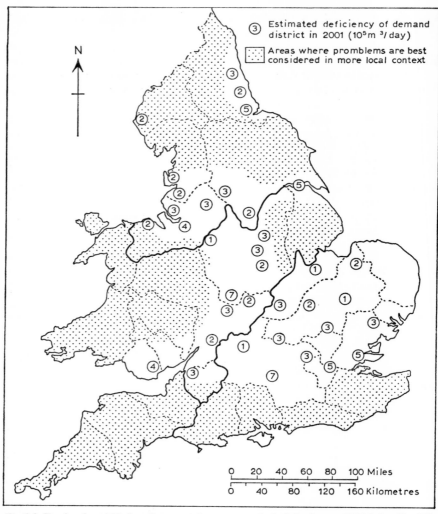

Fig 7.8 England and Wales: forecasted water supply deficits for AD 2001 (after Ryd 2)

HERSHFIELD, D. M. 'Estimating the Probable Maximum Precipitation', *J Hydraulics Division, Proc American Society of Civil Engineers*, **87** (1961), 99–116.
LANGBEIN, W. B. 'Water Research' in *Water Resources Use and Management*, Proc Symp Austral Acad Sc, Canberra, 1963 (Melbourne, 1964).
MINISTRY OF AGRICULTURE, FISHERIES AND FOOD. 'The Calculation of Irrigation Need', *Technical Bulletin*, No 4 (1954).
——. 'Potential Transpiration', *Technical Bulletin*, No 16 (1967).
PENMAN, H. L. 'Natural Evaporation from Open Water, Bare Soil and Grass', *Proc Roy Soc* (A), **193** (1948), 120–48.
ROAD RESEARCH 1969. *Annual Report of the Road Research Laboratory* (HMSO, 1970).
RODDA, J. C. 'A Country-wide Study of Intense Rainfall for the United Kingdom', *J of Hydrology*, **5** (1967), 58–69.
——. 'On the Questions of Rainfall Measurement and Representativeness', *Proc Symp World Water Balance*, IASH/UNESCO, Reading, IASH Pub No 92 (1970), 173–86.
RYDZ, B. 'Regional Water Analysis', *Proc Institution of Civil Engineers*, **49** (1971), 129–43.
SHARP, R. G. 'Estimation of Future Demands of Water Resources in Britain', *J Institution of Water Engineers*, **21** (1967), 232–49.
WATER RESOURCES BOARD. *The Wash: Estuary Storage*, Report on the Desk Study (HMSO, 1970a).
——. *Water Resources in the North* (HMSO, 1970b).
——. *Eighth Annual Report of the Water Resources Board for the Year Ending 30th September, 1971* (HMSO, 1972).
YEVJEVICH, V. 'Misconceptions in Hydrology and their Consequences', *Water Resources Research*, **4** (1968), 225–32.

CHAPTER 8　　　　　　　　　　　　A. K. BISWAS

Application of Mathematical Models to Water Resources Systems Planning

Introduction

Even though the application of modelling techniques to water resources planning and decision-making is of comparatively recent origin, the study and use of models probably antedates recorded history (Orcutt (1960)). Prior to 1950 the use of models by engineers, economists and social scientists engaged in water resources planning and management was quite limited. However, since World War II, there has been a tremendous increase in the use of mathematical modelling as a problem-solving technique, especially in the defence and aerospace-oriented industries, under such labels as 'systems analysis', 'operations research', 'linear, dynamic and geometric programming', 'management science', 'simulation techniques', etc.

The major impetus on the application of systems analysis techniques to water resources planning and management came from the publication of the book *Design of Water Resource Systems* (Maass et al (1962)) by the Harvard Water Program, in 1962. It was an attempt to integrate economic, engineering and governmental planning aspects in designing multi-purpose and multi-unit systems. The Group attempted computer simulation of a simplified river basin, and also developed mathematical programming models for river systems. Since then, considerable progress has taken place in the whole field of modelling techniques as applied to water resources policy planning and decision-making (Biswas and Reynolds (1969, 1970), Biswas (1972)).

Mathematical Modelling

Mathematical modelling is a problem-solving technique wherein attempts are made to build a replica of a real-world system or situation, with the objective of experimenting with the replica to gain some insight into the real-world problem. The system is represented by a series of mathematical expressions in such a way that the resulting relationships accurately describe the phenomenon. The parameters that affect the system are included, along with the factors that influence the parameters. Thus, in a real sense, it is implied that a good mathematical model of any system needs a thorough knowledge and understanding of that system. However, since in most projects involving water resources planning and management, all the factors affecting the system are not known, or if known, several of them cannot be evaluated and quantified, the resulting model does not exactly describe the real-world situation, but if properly developed, it will fairly closely reproduce the critical factors for all practical purposes.

In most cases of mathematical modelling, no attempt is made to include *all* the parameters and variables, otherwise the model will become too unwieldy. The complexity of any model should be dependent on the types of information desired, and its intended use. Often it is more preferable to use a set of ordinary differential equations rather than a set of complex non-linear, partial differential equations. However, the limitations of the assumptions made to simplify a real-world, complex situation, so that it can be modelled and manoeuvred relatively easily, should be clearly understood. In other words, all the models that are constructed are valid over a specified range, and the planners and decision-makers will do well to remember this limitation.

There are six considerations which are basic to the art of model building. They are as follows:

1 the objectives of the system
2 the environment of the system, specifically the constraints
3 the components of the systems, their goals and interrelationships
4 the resources of the system

5 the criteria for measuring the objectives of the system, and the goals of the sub-systems

6 the optimal management of the total system

Biswas and Reynolds (1970) have discussed these aspects in considerable detail elsewhere.

Programming and Descriptive Models

In general, we can divide mathematical models into two categories, programming and descriptive, depending on the relationship of the model to problem-solving. Programming models, for a given objective function, attempt to derive the optimal policy. Descriptive models, on the other hand, attempt to predict possible future consequences due to a set of assumed, exogenous variables and policy alternatives.

Theoretically, water resources managers should find programming models more relevant as an aid to decision-making, since they are geared to obtain optimal policies, directly or indirectly. However, programming models are valid for rather simplified systems which can assume linearity of functional relationships. Often it is not possible to define objective functions, especially in the field of water-resources systems-management, because of the many conceptual and empirical issues associated in constructing a truly comprehensive social-welfare function. Examples of this type of modelling effort are several linear programming models in the water quality management field (Biswas *et al* (1972b), Loucks and Lynn (1964, 1965, 1966, 1967, 1968, 1969 and 1970)), and the interregional, linear programming models of Henderson (1958), Stevens (1958), and the interregional, linear investment model of Rahman (1963).

Objective functions need not be defined for predictive models as they are not directly related to objectives. These types of models predict the values of endogenous variables for a given set of exogenous variables. Exogenous variables can be chosen by the model-builder or the decision-maker, and are often called policy or control variables. If it is assumed that the future consequences are functions of policy variables, the

L

decision-maker can select an 'optimal' policy by changing the policy variables, which will give conditional predictions of future states.

Mera (1969) has pointed out the advantages and disadvantages of this type of modelling efforts in quite succinct terms:

> The predictive model is inferior in the sense that each alternative examined must be chosen subjectively by the persons constructing the model and, therefore, the chance of missing a significant alternative cannot be eliminated. However, it is superior in many senses: the future state can be described in far more detail, the future state can be evaluated differently for different preferences, and the model can be used to test the sensitivity of response to any particular policy variable.

Milliman (1968) and Hamilton et al (1969) have recently presented two comprehensive reviews of these types of models for forecasting regional economic activities.

Decision-making

Mathematical modelling can significantly help decision-makers to arrive at better decisions than would otherwise be possible (a) by broadening his information base, (b) by predicting the consequences of several alternative courses of action, or (c) by selecting a suitable course of action which will accomplish a prescribed result. Hence, modelling provides the relevant facts and alternatives; the decision-maker chooses the strategy. The problem, however, is that in any comprehensive planning of water and other related resources, an infinite number of alternatives present themselves. Admittedly, we can explore more alternatives by mathematical modelling with the aid of current high-speed digital computers than would otherwise be possible, but it should be realised that even utilising the most sophisticated techniques and the latest computers, we cannot examine all the possible alternatives. Such an analysis is neither economic nor practical. For example, if we assume that there are 52 major design variables in a moderately-sized river basin study, as was the case with the Harvard Water Program (Dorfman (1965)), and if each of these variables are assigned only three values, ie, expected, 20 per cent higher and 20 per cent lower, we would have a total of 3^{52} or about 6 million billion billion designs to analyse. Clearly, it is an impossible task. Thus, we have to be selective, and hence, modelling necessitates sampling. Unfortunately, our present state of knowledge of sampling in 52 or more dimensions leaves much to be desired (Biswas (1971)).

Careful analyses of past examples of application of mathematical modelling techniques to water-resources systems-planning do indicate that this technique, in many instances, has given rise to improved decision-making. As our understanding of the real-world situation improves, the modelling techniques become more sophisticated, and as more powerful computers become available, it becomes axiomatic that the models can further improve the process of decision-making.

The mathematical models developed for the comprehensive planning and management of the Saint John River System are briefly described herein.

Mathematical Models for the Saint John River System

The Saint John is an international river, and flows through the state of Maine in the United States and the provinces of Quebec and New Brunswick in Canada. Its length is approximately 700km (435 miles) and has a drainage area of 54,934sq km (21,210sq miles). The comprehensive mathematical models that have been developed apply to the Canadian portion of the river (Fig 8.1). The major industry in the area has traditionally been the harvesting and processing of forest products. The major source of pollution affecting water quality has been the pulp and paper industry. The food-processing industry, especially potato-processing, also contributes significantly to the pollution load. In addition, municipal pollution occurs throughout the length of the river. Fifty-nine significant sources of pollution were identified between the headwaters and Oromocto during the process of development of the models.

The river has been used extensively for logging, and also serves as a

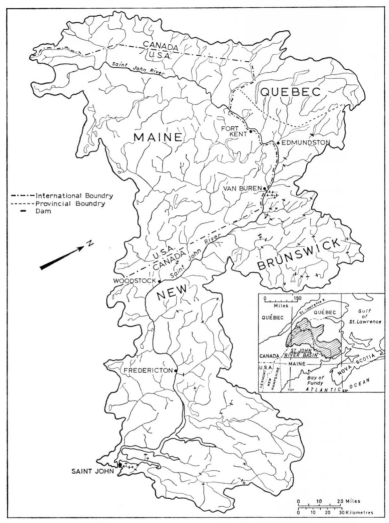

Fig 8.1 General map of the Saint John River basin

transportation link. Its tributaries form the most significant spawning ground for the Atlantic salmon on the east coast of Canada, and consequently, the Saint John River system has been a vital element of the fishing industry for the Maritime Provinces. The river has always been used for recreational purposes, and the fish, wildlife and water fowl in the basin depend on the river and its tributaries for survival. The river flow is partially regulated to develop hydro-electric power, and the current total installed capacity is 530mW.

Two models were developed for this study: a linear programming model and a simulation model. The use of two such models for water resources planning and management is not new. Loucks (1969) suggested the use of programming models for the preliminary screening of alternative designs and operating policies to be followed by simulation models for more detailed analyses of a select few designs and operating policies. After careful consideration of all the factors involved, it was decided to use a similar approach in the present study.

Linear Programming Model

The mathematical programming model developed in this study is a screening device which uses either dissolved oxygen (DO) or biochemical oxygen demand (BOD) as the quality indicator (and BOD as the pollutant). Other types of pollutants can be analysed by the simulation model. The model developed was that of a 'transfer coefficient' type rather than 'explicit reach' type. It is also primarily deterministic and steady-state type of model, stochastic considerations being limited to the selection of appropriate low-flow hydrology.

Essentially, the model consists of two programmes. The first part utilises the available data to generate a set of transfer coefficients and prepares the data in a suitable form for input to the second part, the actual optimisation programme. A detailed report on the sophisticated mathematical techniques developed for the study is available from the author (on request via the Editor), and hence, will not be discussed in this paper.

To develop the models, all the waste sources including industries and municipalities were located and waste loadings to the river were determined. Cost curves were derived for each waste source giving the total annual cost of capital amortisation and operating expenses for a pollution control facility that would remove varying percentages of BOD from the waste effluent. Some of the points on these curves would correspond to primary, secondary and tertiary waste treatment. The model computes the level of water quality resulting in the river from the waste sources, taking into account the assimilative capacity of the river to stabilise oxygen-demanding wastes. The optimising feature of the model involves determining the total minimum cost of treatment in the system, subject to quality constraints for dissolved oxygen levels at various selected points. The optimal solution, then, for a given set of flows in the river and its tributaries is obtained, consisting of a set of industrial and municipal waste treatment plants, each removing the 'optimal' percentage of oxygen-consuming waste from their effluent at minimum total cost.

In addition to DO and BOD constraints, budget constraints can also be added. These budget constraints can be used for specifying:

(a) Minimum treatment levels throughout the basin, or for portions of the basin.
(b) Maximum treatment levels at all points or for portions of the basin.
(c) Total investment in one part of the system to be less than or greater than a certain amount.
(d) Total cost of treatment in one part of the system to be a specific ratio to the total cost of treatment in another part of the system.

These budget constraints represent certain types of policy options and institutional constraints.

A second objective function has been developed which permits maximisation of a 'social quality of water' objective. The approach is to input weight coefficients at each DO quality point which represent the relative social worth of water quality at that point. These subjective coefficients are based on the social willingness to use one region of the river relative to another, and reflect such social parameters as quality of beaches, accessibility, nearness to population centres, etc.

It is interesting to note that most previous applications of a water quality optimisation model have focused on the carbonaceous component of the effluent waste and neglected the nitrogenous component which decays more slowly. The carbonaceous component represents, quite accurately, the total decay behaviour of the waste during its first five days, and is therefore sufficient for rivers with small travel times. However, for the Saint John River, as well as the majority of Canadian rivers, travel times are long and the effects of the nitrogenous component of the effluent discharge become significant. Consequently, the model developed has separated each waste loading into carbonaceous and nitrogenous components, and each treatment plant has been characterised by removal efficiencies for both the carbonaceous and nitrogenous loadings (Biswas *et al* (1972a)).

Some of the policy questions the mathematical programming model can consider are:

(a) What is the change in the level of treatment required and in the cost of pollution control corresponding to an increase or decrease in the overall standard of water quality?

(b) If, in addition to overall standards for water quality in the river, a minimum treatment level is prescribed at every waste discharge point, what is the effect on the cost of treatment at each point, as well as the overall cost of pollution control?

(c) If budget constraints apply to any industry or municipality limiting the degree of treatment that can be provided, how can the standard water quality be achieved or approached at minimum overall cost?

(d) What is the best water quality that can be achieved with overall budget constraints?

(e) Can a reduction of treatment cost be achieved by changing the discharge location of one or more effluents?

(f) What is the effect on treatment costs of an improvement in waste treatment processes?
(g) Which pollution control programme will achieve, over a period of time, the greatest improvement in water quality at minimum social costs? For this purpose the calculations of required treatment installation and treatment costs can be repeated for a number of years using projected future industrial and municipal waste loads. By comparing the optimum configuration of facilities determined for each condition, a suitable programme can be selected.
(h) What is the probable effect on water quality of alternative management procedures and policies such as the application of effluent charges?

The Simulation Model
The simulation model is designed to capture the effects of stream quality under time-varying regimes of river flow and effluent generation. It has also been designed to evaluate effects on stream quality under a variety of river-basin operating conditions, ie, alternative rules of operation for system reservoirs. In addition to examining the effects of biodegradable pollutants on a time-varying basis, it also examines the concentrations of non-degradable pollutants.

In its structure, the simulation model may be conveniently subdivided into three segments:
1 seasonal synthetic hydrology generator
2 daily inflow hydrology generator
3 flow and pollutant routing.

Because of the lack of continuous long-term historical records of river flow, a statistical flow generating programme was developed so that longer time periods of any desired length could be investigated. These river flows were generated based on the statistical characteristics of the existing flow records. Hence, at least conceptually, a synthetic record should be statistically indistinguishable from the historical record. Typical stream-flow records, spanning 30 years (typical for

Canada), do produce acceptable estimates of mean annual and mean seasonal flows and their variances. Therefore, the requirement is confined to a stochastic model that will generate synthetic flow sequences for as long a period of time as required, and which reproduces the historical means and variances in the synthetic sequence, and generates flows with the same distribution as is expected for the total previous history of the river.

The second segment, the daily hydrology inflow generator, generates a set of daily flows while satisfying the synthetic seasonal flows. This is necessary because of the need to examine water quality on a daily basis.

The third segment consists of two distinct phases: flow routing and pollutant routing. The reason for considering these separately is that the changes in actual stream flow are propagated downstream at the wave celerity, while changes in concentration of waste loads are propagated downstream at the average-flow velocity which is generally much slower than the wave celerity.

The model produces a summary print-out of results and detailed output on magnetic tape for further potential analysis, including time history of dissolved oxygen and pollutant concentration levels at all quality points in the system. The vast quantity of data contained on the magnetic tape is summarised in the form of statistics of the frequency of violation of standards, the persistence and magnitude of the violations and the correlation between violation at different points for the standard print-out.

Some of the policy questions the simulation model can answer are:
(a) The variation of water quality is of major concern from a biological point of view, as water quality at certain times of the year is critical for the survival of fish. The problem can be investigated.
(b) The simulation model provides information concerning the severity and longevity of violation of stream-flow standards for a given set of treatment conditions.
(c) The simulation model simulates the flow of eight conservative

pollutants through the river basin, and concentrations at each significant point can be determined.
(d) The model is refined to the point that the effect of reservoirs on water quality can be studied.
(e) The model will indicate, in far more detail than the linear programming model, the benefits, in the sense of improved water quality, to be derived from specific investments in waste treatment facilities. It could therefore be used to confirm or refine investment recommendations based on the results of the linear programming model.
(f) The model provides opportunities for assessing the effect of stream-flow regulations and augmentation on water quality, as well as the effect of waste reduction at selected critical periods, and the effect of instream treatment.

Both models are rather flexible and can be applied to other river basin configurations. The simulation model can be adjusted using a wide range of options that can control the simulation period, output format, characteristic time intervals, system-operating parameters and all the relative design variables.

Conclusion

The application of mathematical modelling to water-resources policy-planning and decision-making is not new; what *is* new, however, is our present capability to analyse more complex systems. The ideal model, from the decision-maker's point of view, is the one that can replicate all the essential structural and decision-making elements of the real-world system under consideration. The more the models can be manipulated, the greater is their usefulness to policy-planners and decision-makers in exploring the possible impacts of alternative policies.

Decision-making in the field of water resources management is incremental rather than total. The decision-makers proceed through one stage at a time, comparing the consequences of different feasible policies. Thus, the interaction between the decision-makers and the models promises the greatest possible future benefit, especially in terms of 'optimal' policies. Models cannot replace experience; in fact they augment it.

References

BISWAS, ASIT K. 'Mathematical Models and Their Use in Water Resources Decision-Making', *Proceedings, 14th Congress International Association for Hydraulic Research*, **5** (Paris, 1971), 241–8.

——. *Systems Approach to Water Management* (New York, 1972).

BISWAS, ASIT K., DURIE, R. W. and REYNOLDS, P. J. 'Water Resources Systems Planning', *Symposium Proceedings, 8th Congress, International Commission on Irrigation and Drainage* (Varna, Bulgaria, 1972), S117–S134.

BISWAS, ASIT, PENTLAND, R. L. and REYNOLDS, P. J. 'Water Quality Modelling: State-of-the-Art', in Biswas, Asit K. (ed), *Proceedings, International Symposium on Mathematical Modelling Techniques in Water Resources Systems* (Ottawa, 1972), 481–96.

BISWAS, ASIT K. and REYNOLDS, P. J. 'Socio-Economic Simulation for Water Resource System Planning', *Proceedings, 13th Congress, International Association for Hydraulic Research*, **1** (Kyoto, Japan, 1969), 75–82.

——. 'Hydro-Economic Models for Water Resource System Planning', *Discussion Paper 70–4*, Policy and Planning Branch, Department of Energy, Mines and Resources (Ottawa, 1970).

DORFMAN, R. *Formal Models in the Design of Water Resources Systems*, Water Resources Research, **1**, No 3 (1965).

HAMILTON, H. R., GOLDSTONE, S. E., MILLIMAN, J. W., PUGH, A. L., ROBERTS, E. B. and ZELLNER, A. *Systems Simulation for Regional Analysis: An Application to River Basin Planning* (Cambridge, Massachusetts, 1969).

HENDERSON, J. M. *The Efficiency of Coal Industry* (Cambridge, Massachusetts, 1958).

LOUCKS, D. P. *Stochastic Methods for Analyzing River Basin Systems*, Cornell University Water Resources and Marine Sciences Center, Technical Report No 16 (Ithaca, 1969).

LOUCKS, D. P. and LYNN, W. R. 'A Review of the Literature on Waste Water and Water Pollution Control: Systems Analysis', *J of the*

Water Pollution Control Federation, **36**, No 7 (1964); **37**, No 7 (1965); **38**, No 7 (1966); **39**, No 7 (1967); **40**, No 6 (1968); **41**, No 6 (1969); **42**, No 6 (1970).

MAASS, A., HUFSCHMIDT, M. M., DORFMAN, R., THOMAS, H. A., MARGLIN, S. A. and FAIR, G. M. *Design of Water Resource Systems* (Cambridge, Massachusetts, 1962).

MERA, K. 'Survey of Model Building for Regional Economics', *Discussion Paper No 55*, Program on Regional and Urban Economics, Harvard University (Cambridge, Massachusetts, 1969).

MILLIMAN, J. W. *Large Scale Models for Forecasting Regional Economic Activity: A Survey*, School of Business, Indiana University (Bloomington, 1968).

ORCUTT, G. H. 'Simulation of Economic Systems', *The American Economics Review*, **50**, No 5 (1960), 893–907.

RAHMAN, M. A. 'Regional Allocation of Investment', *Quarterly J of Economics*, **77** (1963).

STEVENS, B. H. 'An Interregional Linear Programming Model', *J of Regional Science*, **1** (1958), 60–98.

CHAPTER 9 I. BURTON, D. BILLINGSLEY, M. BLACKSELL, V. CHAPMAN, A. V. KIRKBY, L. FOSTER and G. WALL

Public Response to a Successful Air Pollution Control Programme

The smoke control programme embodied in the British Clean Air Act of 1956 was one of the earliest attempts to control atmospheric pollution by government action on a national scale. It has been widely described as successful, and is frequently cited as an example to support arguments that pollution can be controlled without resort to harsh legislation and severe penalties.

The time that has elapsed since the passage of the Act in 1956 affords an opportunity for a retrospective analysis of its accomplishments. This paper reports briefly and selectively on a study undertaken by a group of geographers, drawing in part on sample interview data from Edinburgh, Exeter, and Sheffield in the United Kingdom, and from papers on special aspects of air pollution prepared by members of the group. Focus is on (a) how the Act has worked; (b) how air pollution is now perceived by the general public; and (c) what people consider doing in response to the threat of air pollution. Policy implications of the findings are discussed; a more complete statement of these is expected to be published elsewhere.

The Successful Clean Air Act, 1956
All discussion of air pollution in the United Kingdom takes place against the background of the Clean Air Act, 1956 (4 and 5 Elizabeth 2,

Ch 52) which has been credited with dramatic reduction in smoke concentration occurring in London and other large cities. The First Report of the Royal Commission on Environmental Pollution (1971, para 36, p 11) states: 'Since the first Clean Air Act became law in 1956 there has been a steady reduction in the emission of smoke and sulphur dioxide into the air over Britain . . .' and warns that 'The downward trends in smoke and sulphur dioxide pollution are encouraging, but they will continue *only if* there is no relaxation in applying the provisions of the Clean Air Acts and the Alkali Etc. Works Regulation Act' (para 38, p 12, italics added). The commission report also includes a graph (Fig 9.1) which shows the decline in average smoke concentrations from 1958 to 1968. The argument that this decline is due to the

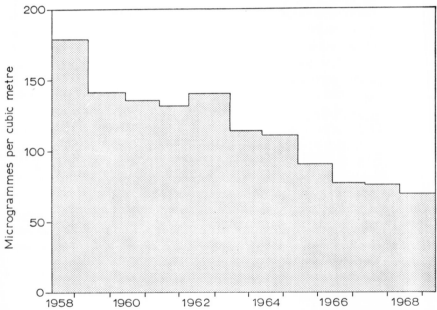

Fig 9.1 Average smoke concentration near ground level in the United Kingdom (1958–68)

operation of the Clean Air Act rests on rather slender evidence. It has been shown elsewhere (Auliciems and Burton (1972)) that the average smoke concentrations at Kew have been declining since records were first kept in 1922–3. Fig 9.2 shows a line fitted to the data from 1922–3 to 1955–6 and extended to 1970–1, suggesting that recorded smoke reductions since 1956 might well have occurred in the absence of the Clean Air Act.

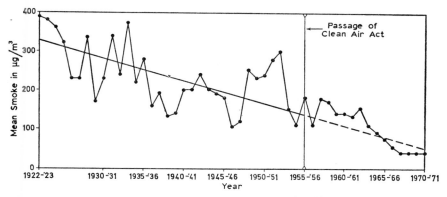

Fig 9.2 Average smoke levels at Kew, October–March, 1922–71

That factors other than legislation have contributed to the decline in smoke concentrations is recognised in part by a Warren Spring Report which states:

> This continued replacement of coal by oil [in industry] suggests that it is reasonable to credit the Act with having greatly accelerated developments which would, however, of themselves, have taken place slowly but inevitably. In fact, like other successful legislation the Clean Air Act was 'swimming with the tide' of industrial development. (Craxford, Gooriah and Weatherly, 1970.)

Domestic heating practices were also changing prior to the Act, and especially in the more affluent areas, different fuels were substituted for coal. Other reasons for reduction of smoke concentrations include (a) industrial conversions to oil, electricity and gas, (b) electrification and dieselisation of the railways, (c) urban renewal and redevelopment schemes, as well as (d) the conversion of domestic heating from coal fires to smokeless fuels.

Smoke Control Areas

The main target of the 1956 Act was smoke, especially that emitted from domestic chimneys. In so far as the Act has been instrumental in smoke abatement, it is largely through its success in bringing about a change in domestic heating practices away from the traditional coal fire to the use of 'smokeless fuels' or gas or electricity. Smokeless fuels include coke, coalite, or other similar manufactured fuel products. The attachment of the British to the coal fire is legendary, and few would have predicted that in such a 'hearth-centred culture' the substitution of other fuels, and the conversion to gas and electric heating appliances, could take place so rapidly and with such apparent ease. What is more remarkable is that the change has largely been accomplished by permissive legislation which *enables rather than compels* local governments to take action. However, after twelve years of successful operation of the Act, the Central Government took the power to *require* local government action, although this has not yet been exercised (Clean Air Act, 1968, Section 8).

The process of smoke abatement in Britain is the establishment by the local authorities of smoke control areas within which the emission of smoke from a chimney is made an offence under the Act. The local authority is required to repay 70 per cent of the reasonable cost of conversions necessary for homeowners to comply with the smoke control order. Of this amount, $\frac{4}{7}$ may be reclaimed from the Central Government by the local authority. This cost-sharing arrangement, whereby the Central Government pays 40 per cent, the local authority 30 per cent, and the property owner 30 per cent of the cost of conversions, seems to have worked satisfactorily in some areas, but not all.

Local Authority and Public Response

Local authority response is currently being examined in some detail (Foster (1972)). The aggregate pattern for 'black areas' is illustrated in Fig 9.3. The 1956 Act was followed by a rapid rise in the number of local authorities adopting smoke control programmes. In recent years this has levelled off. Most authorities in polluted areas (the specially designated 'black areas') have adopted programmes although some have still failed to do so. The initial adoption of a programme does not by itself guarantee rapid progress in smoke abatement thereafter. Table 9.1 shows the proportion of the area covered by smoke control orders is quite variable. On a regional basis it is highest in London and generally lowest in the coalfield areas, especially in the East Midlands and the Northern region. (Wales is a special case due to the higher quality coal which produces less smoke, but the south Wales black area also happens to be on a major coalfield.) Clearly the process still has a long way to go before all the black areas become smokeless.

Table 9.1

VARIATIONS IN THE ADOPTION OF SMOKE CONTROL AREAS

Region	Acreage covered* %
Northern	31·9
Yorks and Humberside	49·6
East Midlands	24·3
West Midlands	34·4
Greater London	73·8
North Western	49·2
South Western	28·5
Wales	0·01

* Black area authorities only, as of 31 December 1970.
Source: Foster (1972), 12

In general, larger authorities seem to be more willing to adopt smoke control programmes than smaller authorities. Also, wealthier authori-

178 CLIMATIC RESOURCES AND ECONOMIC ACTIVITY

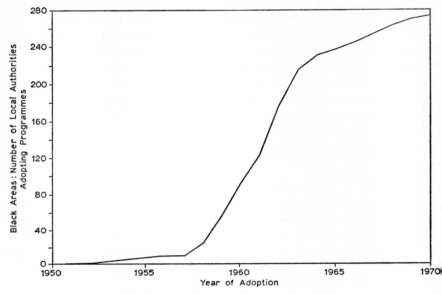

Fig 9.3 Adoption of smoke control programmes

ties, in terms of rateable value per capita, have made most rapid progress in acreage covered. Presumably the richer areas of the country can better afford the cost of conversions, both those borne by local authorities and by property owners. It has also been suggested that the larger authorities are more likely to have the technical staff and expertise available for the development of a phased programme. Other factors are probably at work also, among them the more active role of people in the wealthier communities in seeking solutions to their environmental problems.

Resistance to the adoption of smoke control programmes is difficult to gauge accurately. What strong opposition there has been tends to come from coalfield communities where cultural and economic attachment to coal is strong and where coal miners are entitled to a certain amount of free or concessionary coal. Another locality where strong resistance has developed is in the London borough of Epsom and Ewell. This is the only local authority designated as a 'black area' within Greater London which has yet to adopt a smoke control programme. It refuses to do so on the grounds that it is not badly polluted, and should not have been designated as a 'black area'.

Among the overwhelming number of local authorities which have adopted smoke control programmes, some have moved ahead rapidly while others have proceeded cautiously. It is not known to what extent this may be due to real or supposed public opposition and fear of loss of political support. Clearly, in some areas, to require residents to pay 30 per cent of the reasonable cost of converting their fireplace would be unpopular, especially among those with low or fixed incomes, and the old. Such groups are proportionately more numerous in the older and higher-density residential districts found closer to the centre of a city, and adjacent to or scattered within the industrial zones. In areas of this type, pollution levels from both domestic, and especially industrial, sources are generally higher. Evidence from interviews, however, suggests that the populations of such areas are generally more passive and less likely to complain (Kirkby (1972)). A somewhat stronger reason is that central areas contain more institutional users of coal, including older industrial establishments, where the costs of conversion would be significantly higher.

Smoke control areas are often first designated in areas of new housing which can more easily build in smokeless fuel facilities. It is also usual to begin the designation of smoke control areas in suburbs on the western side of a city, eg for Edinburgh (Billingsley (1972)) where advantage was taken of the direction of the predominant winds, thus allowing a slow approach to areas of highest pollution.

A similar pattern may be observed in Sheffield (Wall (1972)) where early designations included large areas of virtually unoccupied hill and moorland on the western side of the city. The inclusion of open country and low-density modern housing estates in the first smoke control areas creates an exaggerated impression of the progress of smoke control, since both the number of acres and premises within the designated areas are proportionately larger than the percentage reduction of smoke emissions. Nevertheless, the policy makes good practical, as well as political sense. Smoke control programmes are established first in the most receptive areas and their value is thus demonstrated to others. Also, the longer the delay, the lower the cost to the local authority. In the tenements typical of central Edinburgh, for example, voluntary conversions to gas, oil, and electricity are occurring anyway. In Sheffield much of the terrace housing in one of the polluted areas in Carbrook (Attercliffe) is being removed under urban redevelopment schemes. By proceeding slowly, local authorities lessen the amount of expenditure necessary to extend control into the more densely populated and more highly polluted core areas.

In addition to the so-called 'black areas' deemed to be priority cases for pollution abatement, some non-black areas have also adopted smoke control programmes. As in the case of Exeter, pollution levels tend to be relatively low, and smoke control is adopted for a variety of idiosyncratic reasons. In Exeter the progress of smoke control has largely taken place at the insistence of an active medical officer of health who is concerned about the effects of *indoor* air pollution from coal fires (Blacksell (1972)).

It is difficult to reconstruct accurately the pattern of public response to the introduction of smoke control. In Sheffield there has been little opposition (Wall (1972)). Pressure groups working for cleaner air have been in existence in the city since before the Clean Air Act, and it was felt by some that the Act did not go far enough since industrial emissions were left under the relatively weaker control of the Alkali Inspectorate of the Central Government. Public opposition to smoke control orders seems to have been somewhat stronger in Edinburgh (Billingsley (1972)) and objections more numerous. None has succeeded in blocking an order, however. The local authority has responded by taking a tolerant view of infractions.

In Exeter, a city which had substantially less pollution in 1956 than

Sheffield or Edinburgh, the overwhelming public reaction to smoke control was adverse (Blacksell (1972)). Criticism came from the Exeter and District Coal Traders' Association which argued rather strongly that the livelihood of its members was jeopardised. The general public also complained in letters to the press that since Exeter was already a healthy and clean city, smoke control was unnecessary and would impose an additional burden of higher property rates and require people to buy the more expensive smokeless fuel. However, smoke control was adopted and there have been no significant complaints in recent years.

Awareness and Evaluation

From questionnaire survey data collected in Edinburgh, Sheffield, and Exeter in 1971, it is possible to describe with greater accuracy the present state of public perception and understanding. In each city a stratified random sample was selected to include respondents in roughly equal proportions from areas of relatively high, medium, and low pollution levels. The National Survey of Air Pollution data from Warren Spring Laboratory show that average winter smoke concentrations in Exeter ranged around $40 \mu g/m^3$ in the period 1966–9. Sheffield and Edinburgh values were approximately equal and ranged around $120 \mu g/m^3$.

In Exeter and Sheffield (Table 9.2) a large majority of people interviewed denied the existence of an air pollution problem in their neighbourhood.

Table 9.2
AWARENESS OF AIR POLLUTION

Question: Is there an air pollution problem in this area?

	Exeter		Sheffield		Edinburgh	
Yes	26	22%	30	25%	190	53%
No	86	72%	78	65%	166	47%
Doubtful	4	3%	8	7%	–	–
Don't know	4	3%	4	3%	–	–
TOTAL	120	100%	120	100%	356	100%

In Exeter this viewpoint has been strong for many years as shown by the initial opposition to smoke control. In Sheffield it is relatively new. The population of Sheffield has changed its perception of air pollution. It used to be thought that Sheffield was a very smoky and dirty city. Now many people believe the pollution problem has been entirely or very largely solved. This is due only in part to the dramatic improvement that has occurred in smoke levels. Undoubtedly, the active publicity campaign launched by the city, in which it claims to be 'the cleanest industrial city in Europe', has had a significant impact on the image that Sheffield residents have of their city (Kirkby (1972)). In some localities pollution still reaches high levels. At Surrey Street, for example, the highest day value of smoke concentration in 1969–70 was 605μg/m^3, although the mean winter value was only 69 (Gooriah and Weatherly (1971)).

In Edinburgh a significantly larger proportion of the respondents, 53 per cent, recognise the existence of an air pollution problem. Although Edinburgh levels do not differ greatly from those now prevailing in Sheffield, there has been slower improvement in Edinburgh and there is no equivalent of the Sheffield publicity campaign.

The general pattern that emerges is that many people in the selected UK cities studied do not currently regard air pollution as a serious problem. In the absence of comparable data for earlier years it is not possible to state with confidence what the trends have been. A plausible hypothesis is that there has been some public awareness of air pollution for a long time, at least since the London smog episode of 1952, but that this awareness has not been translated in Britain, as it has elsewhere, into a high degree of public anxiety.

The growing social and political concern for environmental quality, through the 1960s, in most western industrial countries, has usually included air pollution (Swan (1972)). One recent study in Toronto found that air pollution was considered the most important urban problem (Auliciems and Burton (1970)). This contrasts with the results of a study of public and professional evaluations of environmental problems in the East Riding of Yorkshire (Chapman (1972)). Respondents were asked to describe each of 14 environmental problems as (1) not serious, (2) slightly serious, (3) moderately serious, (4) very serious, or (5) extremely serious. In the ranking of the scores derived from 200 general public respondents, air pollution was ranked 11th out of 14. Responses from a sample of 90 professional personnel concerned with environmental issues as part of their job (government officials, in-

Table 9.3

PERCEIVED SERIOUSNESS OF ENVIRONMENTAL PROBLEMS BY THE GENERAL PUBLIC AND BY PROFESSIONAL SAMPLE

	General public sample			Professional sample	
Rank	Problem	Likert scale score	Rank	Problem	Likert scale score
1	Destruction of wildlife	307	1	Litter	294
2	Litter	306	2	Recreation pressures in the countryside	271
3	Over use of pesticides	290	3	Removal of hedges	259
4	Sea pollution	257	4	Improper treatment of sewage	254
5	Recreation pressures in the countryside	252	5	Ex-urban development	247
6	Noise	247	5	Destruction of wildlife	247
7	Spoiling of the landscape	241	7	Water pollution	244
8	Derelict land	240	8	Over use of pesticides	236
9	Removal of hedges	238	9	Sea pollution	227
10	Improper treatment of sewage	231	10	Spoiling of the landscape	225
11	*Air pollution*	224	11	Noise	220
12	Ex-urban development	223	12	Derelict land	209
13	Water pollution	211	13	*Air pollution*	201
14	Deterioration of soil quality	185	14	Deterioration of soil quality	178

Source: Chapman (1972), 8.

dustrialists, environmentalists, and legislators) ranked air pollution even lower, at 13th out of 14 (Table 9.3).

The reasons for an apparent lack of public concern about air pollution in Britain are hard to establish, but circumstantial evidence suggests two strong and mutually supporting explanations. There has been a substantial decline in smoke concentrations in London, Sheffield, and other cities. The single, most visible component of air pollution has noticeably declined. Further, this decline has been widely reported and acclaimed through the media and in public speeches for several years. When our own direct sense-experience bears out what we are repeatedly told, there is a strong inclination towards acceptance and belief. It is not surprising, therefore, that people in Britain are convinced that the air pollution problem has been largely solved or is well on its way towards solution. This is supported also by the widespread acceptance of smoke control areas. In Edinburgh, Sheffield, and Exeter the overwhelming majority now agree that smoke control areas are the best method of fighting air pollution (Table 9.4). In the absence of the Clean Air Act, 1956, the decline in smoke concentrations and the reassuring media coverage, it seems likely that concern about air pollution would have risen as sharply in Britain as in North America.

Table 9.4

EVALUATION OF SMOKE CONTROL AREAS

Question: Are smoke control areas the best method of fighting air pollution?

	Exeter		Sheffield		Edinburgh	
Yes	100	83%	105	88%	127	70%
No	7	6%	3	2%	21	12%
Doubtful	10	8%	10	8%	16	9%
Don't know	3	3%	2	2%	16	9%
TOTAL	120	100%	120	100%	180	100%

Perception and Adjustment

The psychological dimension of internal-external control refers to the degree to which people believe events in their own lives are controlled by themselves or by external forces (fate, God, chance, or other people). A sentence completion test was employed in Exeter, Sheffield, and Edinburgh, and analysed by Kirkby (1972), to examine the relationships between people's response to air pollution (their emotional responses and the type of action they see themselves taking) and internal-external control. The analysis suggests that most individuals are probably not conscious of the air pollution problem until they are reminded about it. When told that government measures such as the Clean Air Acts are successfully being implemented, most people are content to assume that individual action is not necessary.

The most frequent emotional response to air pollution (36 per cent) is concern with physical health symptoms which ranged from generally feeling unwell to specific fears of serious health effects. Anxiety and emotional stress were the second most common response (25 per cent) with a further 13 per cent admitting to feelings of anger, disgust, and resentment.

A small majority of respondents do not expect a bad air pollution situation in the future (Table 9.5). Among the areas sampled, expecta-

Table 9.5

EXPECTATION OF FUTURE EPISODES OF POLLUTION

Question: Do you think there will be a bad air pollution situation in the future?

	Exeter		Sheffield		Edinburgh	
Yes	35	29%	26	22%	48	27%
No	61	51%	63	53%	77	43%
Uncertain	13	11%	21	17%	31	17%
Don't know	11	9%	10	8%	24	13%
TOTAL	120	100%	120	100%	180	100%

tion of future pollution is highest in the Carbrook area of Sheffield (where pollution levels are also highest) and here 40 per cent expect a bad situation in the future (Wall (1972)). A common pattern of response in this area is for people to achieve cognitive consistency by asserting that 'air pollution doesn't bother them', they are 'used to it', they 'don't notice it', or it 'isn't really serious anyway'.

When asked what a person can do in a particularly bad air pollution situation, a common response is to treat the question very lightly. 'What *can* you do?' 'Stop breathing?' On reflection, respondents tend to suggest protective action such as wearing a smog mask (or other personal air filter), staying indoors, and keeping the windows closed (Table 9.6). Very few are disposed to complain; only a small minority

Table 9.6

PERCEIVED ADJUSTMENTS TO POLLUTION

Question: When air pollution is particularly bad, what can a person do?

	Exeter		Sheffield		Edinburgh	
Wear a smog mask or a personal air filter	39	22%	36	23%	45	25%
Stay indoors	38	21%	36	23%	35	20%
Close or seal windows	27	15%	22	14%	16	9%
Complain	26	15%	31	20%	35	19%
Actions to stop or reduce individual emissions	19	10%	11	7%	13	7%
Move away (temporarily)	5	3%	8	5%	13	7%
Other	25	14%	13	8%	24	13%
TOTAL	179	100%	157	100%	181	100%

appear to recognise their own potential contribution to the problem by reducing individual or family emissions of air pollution. There are significant differences in adjustments perceived, however, by age, sex, socio-economic status and experience of the problem. Those who respond more actively to air pollution tend to be men more than women, younger (under 65 years), of higher socio-economic status, and living *outside* the most highly polluted districts. Conversely, more passive responses tend to come from women rather than men, from older people (on fixed incomes), and those living in the areas of highest pollution concentrations (Kirkby (1972)).

Policy Considerations

The evident success of the Clean Air Act, 1956, has led some to conclude that the British model provides a good example for others to follow (Scorer (1961)). It can be characterised as the bureaucratic fix. It has not placed heavy reliance on the development of new technology to solve the pollution problem in the tech-fix manner often used in North America, nor has it made much use of penalties and administrative fines as a means of backing up and enforcing the law as suggested in President Nixon's message to the US Congress on 8 February 1971 when he stated that 'violations of standards and abatement orders be made subject to court-imposed fines of up to $25,000 per day and up to $50,000 per day for repeated violations...'.

British strategy relies heavily on 'reasonableness', on the willingness of people to make compromises rather than on coercive penalties. Above all it assumes a relationship of trust and mutual confidence between central and local government and the general public. Where such relationships cannot be shown to exist, there is little prospect for the successful transfer of the British strategy to other countries.

A question remains, however, concerning the validity of the strategy adopted in Britain itself. Has the Clean Air Act worked as well in reality as it is commonly believed, and might there have been some alternative or additional approaches that would have succeeded as well or better? Such a question is of more than academic interest. There are other countries where the air pollution problem is not unlike that of Britain, including modern industrial nations where coal is still an important fuel for domestic heating and industrial purposes, for example, Denmark, Holland, Hungary, and New Zealand. Furthermore, there are nations in the tropical world that are beginning to confront air pollution problems in their major cities which are analagous to the problem of smoke in the UK. During the winter (cool) season in Calcutta evening temperature inversions concentrate the smoke from countless small fires used for cooking and warmth. The fuels used are low-quality coal and dung cakes made from cow manure. When burnt, the resulting smoke mixed with automobile exhausts and the other emissions of a crowded tropical industrial city produces an atmosphere best described as 'smung'.

A critical appraisal of British experience in curbing emissions of smoke from domestic sources is therefore of potential interest in a number of other countries. Some of the conclusions that emerge from the study reported here have policy implications which necessarily depend on special characteristics of the pollution problem and circumstances obtaining in other jurisdictions.

The Strategy of 'Permissive Legislation'
Letting the local authorities go at their own speed has been most successful in the Greater London area where an added impetus has been given by the severe air pollution episodes of 1952 and 1962. London is also the wealthiest 'black area' and has no retarding effect associated with coal mining areas. The permissive strategy has also worked well in other large and relatively affluent communities and in those where pollution was less serious. It has encountered opposition in some of the northern cities of the UK, and the coalfield areas where the pollution problem remains serious.

Generally, the national and local governments have moved in a cautious and authoritative way to remove smoke from the atmosphere, and costs have been allocated among the population where they can best be borne. Initial adoption of smoke control in the most receptive areas has helped to gain public acceptance for a programme that might well have engendered much more resistance if pressed vigorously in the communities where pollution is higher and opposition stronger. At the local level, smoke control has tended to be adopted first and foremost in those sections of the city where the higher-income, better-educated, and younger people live. Improvements have come more slowly in the areas occupied by those with lower income, less education, and greater age.

Public Response
A generally confident feeling exists in which people do not consider themselves to be seriously threatened by air pollution now or in the future. A commonly held opinion is that pollution is largely a governmental matter and there is little recognition of need for individual action. The most preferred adjustments to a serious air pollution situation are in the direction of self-protection. The public is convinced that individual adjustments (eg conversion of heating systems away from the coal fire) are desirable and necessary; this belief has been created by careful bureaucratic and political leadership.

Qualifications and Limitations
Perhaps the most serious qualification that can be stated about the successful air pollution control programme in the UK is that it has been too obviously successful. The most readily observable pollutant, smoke, has been dramatically reduced in some places and it is lower almost everywhere to some degree. Emphasis on smoke has been accompanied by a relative neglect of sulphur dioxide, oxides of nitrogen, and carbon, lead, and other less well recognised hazards. Information on most of these pollutants is very limited and the available data do not permit confident assessments of the situation. Circumstantial evidence, however, suggests that some of them at least have been increasing and that new air pollution hazards may be emerging. There is very little recognition of the possibility among the public or in government circles. A mood of self-congratulation prevails and credit is claimed for the reduction of smoke concentrations through the use of an enlightened legislation.

The newer forms of atmospheric pollution (being largely invisible, odourless and tasteless) may not prove readily amenable to easily designed control legislation and are less likely to generate public support and co-operation from large corporate organisations for such programmes. Whether air quality management can therefore continue to be successful in the UK without resort to stricter measures, more severe penalties, and a redirection of the pattern of economic growth, remains an open question.

Acknowledgement
The research reported in this paper has been supported by a grant from 'Resources for the Future, Inc', Washington DC, USA.

References
AULICIEMS, A. and BURTON, I. *Perception of Air Pollution in Toronto*, Natural Hazard Research Working Paper No 13. Toronto: Department of Geography, University of Toronto (1970).
——. *Trends in Smoke Concentrations Before and After the Clean Air Act of 1956* (in the press).
BILLINGSLEY, D. *Awareness of Air Pollution in Edinburgh: Private Perception and Public Policy*, unpublished paper prepared for the Man–Environment Commission Symposium of the International Geographical Union (Calgary, 23–31 July 1972).
BLACKSELL, M. *Attitudes towards Smoke Control in Exeter*, ibid (1972).
CHAPMAN, V. *The Public Perception of Air Pollution as an Environmental Problem in the East Riding of Yorkshire*, ibid (1972).
CRAXFORD, S. R., GOORIAH, B. D. and WEATHERLY, M-L. P. M. 'Air Pollution in Urban Areas in the United Kingdom: Present Position and Recent National and Regional Trends', *National Survey of Air Pollution*, Warren Spring Laboratory (Stevenage, 1970), SCCB 74/6.
DEPARTMENT OF TRADE AND INDUSTRY, National Survey of Air Pollution. *The Investigation of Air Pollution*, Warren Spring Laboratory (Stevenage, various dates).
FOSTER, L. *The Adoption of Smoke Control Areas in the United Kingdom*, unpublished paper prepared for the Man–Environment Commission Symposium of the International Geographical Union (Calgary, 23–31 July 1972).
GOORIAH, B. D. and WEATHERLY, M-L. P. M. *National Survey of Smoke*

and *Sulphur Dioxide: South Yorkshire*, Warren Spring Laboratory (Stevenage, 1971).

KIRKBY, A. V. *Perception of Air Pollution as a Hazard and Individual Adjustment to It in Exeter, Sheffield and Edinburgh*, unpublished paper prepared for the Man–Environment Commission Symposium of the International Geographical Union (Calgary, 23–31 July 1972).

ROYAL COMMISSION ON ENVIRONMENTAL POLLUTION. *First Report* (HMSO, 1971), Cmnd 4585.

SCORER, R. S. 'New Attitudes to Air Pollution—the Technical Basis of Control', *Atmospheric Environment*, V (1961), 903–34.

SWAN, J. A. 'Public Response to Air Pollution', in Wohlwill, J. F. and Carson, D. H. (eds), *Environment and the Social Sciences: Perspectives and Applications* (Washington DC, 1972), 66–74.

WALL, G. *Public Response to Air Pollution in Sheffield*, unpublished paper prepared for the Man–Environment Commission Symposium of the International Geographical Union (Calgary, 23–31 July 1972).

CHAPTER 10   K. TAYLOR

Some Aspects of the Economics of Air Pollution in the United Kingdom

Introduction

In June 1969 the Programmes Analysis Unit was commissioned by the then Ministry of Technology to carry out a study on air pollution in the UK. The terms of reference of the study were:
(1) To define sources and types of pollutants currently thought to be important and to examine the changes likely to arise from future technological, social and economic developments.
(2) To assess the relevance of current R & D programmes and to suggest, if possible, in the light of the above, the appropriate directions and levels of further research needed to assist in minimising the total economic and social costs of pollution.

In this paper, a brief description will first be given of the scope of the entire study and the methodology adopted. Secondly, the agricultural and climatic aspects of air pollution will be described qualitatively.

Theoretical Model

The relationship between the cost of damage and the cost of control of pollution can be represented in the manner shown in Fig 10.1.

Suppose that the current level of pollution is X, corresponding to a total cost (damage plus control) C_X, then the maximum net benefit is achieved by reducing the pollution level to Y, where the marginal rate of change of damage cost with pollution level is equal to the

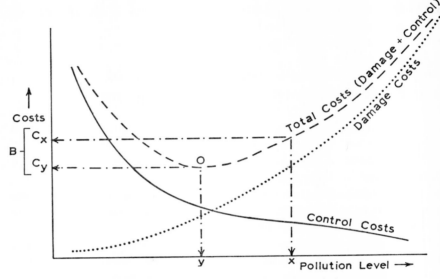

Fig 10.1 Relationship between pollution damage and control costs

marginal rate of change of control cost. The maximum net benefit D is equal to $(C_X - C_Y)$. The important principle to emerge from this model is that it is worthwhile reducing the pollution level only so long as the marginal increase in cost of control is less than the marginal decrease in damage cost achieved thereby.

The question which immediately follows is: what do we mean by cost? Control costs may be defined as those costs incurred in reducing the damage caused by pollution. This may mean a reduction in the quantities of pollutants emitted, or a change in the method of emission (eg by the use of high stacks) designed to reduce the damaging effect for a given quantity of pollutant emitted. Damage costs on the other hand are those costs which are directly or indirectly incurred as a result of pollutants being present in the atmosphere. Although this may appear straightforward, there are a number of difficulties which immediately

become apparent when any attempt is made to assess damage costs in practice. The principal difficulties are:
   (a) The attribution of damage to pollution as opposed to other possible causes.
   (b) The distinction between damage resulting from previous levels of pollution and current levels.
   (c) The distinction between damage manifested as increased expenditure (economic costs), and damage manifested in a decreased quality of life (social costs).

Though none of these is easy to overcome, the distinction between economic and social costs is the most intractable. Often the distinction is not easy to make, but always there is great difficulty in assessing social costs in monetary terms, the most notorious being the problem of evaluating the monetary worth of a human life. It is this question of economic cost versus social cost which bedevils any attempt to assess the damage caused by a public disutility like air pollution.

Damage Assessment

The damage caused (potentially at least) by air pollution was classified under the following heads:
   (1) Increased painting and decorating costs.
   (2) Increased cleaning costs (household, laundry, buildings).
   (3) Corrosion damage to metals and other structural materials.
   (4) Damage to textiles, paper, etc.
   (5) Damage to agricultural produce.
   (6) Damage to health (respiratory system).
   (7) Loss of amenity, reduction in land and property values.
   (8) Damage resulting from pollution effects on climate.

Because of the huge complexities involved in treating each damage incident separately, the crude assumption was made that the UK could be divided into two types of region—polluted and clean. An examination of the average levels of pollution showed that the conurbations, in which one-third of the population lives, could be regarded as having overall significantly higher levels of pollution than non-conurbations, and for most areas of damage this seemed to be the relevant division to make.* However, for agricultural damage, a more appropriate division seemed to be between agricultural areas (a) in the neighbourhood of conurbations, and (b) remote from same.

The pollutants of principal concern in the study are sulphur dioxide and particulate matter, this being a matter of necessity since the only existing large-scale data available relate to these two materials. Nevertheless, whenever possible, other pollutants, such as carbon monoxide and oxides of nitrogen, have been taken into consideration.

Firstly, the incremental damage costs were calculated. These are the excess costs incurred in the polluted areas over and above those incurred in the clean areas. Secondly, by making certain assumptions about the likely general shape of the damage cost/pollution level curve, the total costs were estimated, which included the costs in both polluted and clean areas down to threshold pollution level. With reference to painting, etc, cleaning, and textile damage, it was argued that the real cost of damage was zero at the current levels of pollution in clean areas since below these levels the lives of the affected materials would be determined by factors other than pollution, eg fashion or taste. Thus, in these instances, the total damage cost equals the incremental damage cost. On the other hand, for corrosion, health and agriculture, it was believed that the threshold level was more likely to be at zero pollution level.

Future total costs in 1980 were then estimated using expected future trends in pollution emissions. This date was chosen as being the earliest time by which any recommendations arising from currently initiated research and development would be put into widespread use. It appeared that by this date the damage costs in the areas of painting, cleaning, and textiles would be very small compared with those associated with corrosion, agriculture, and health.

* The impact of the Clean Air Act on London has been so marked that it is included in the non-conurbation category. (See Chapter 9, herein.)

Allocation of Damage Costs Between Sources

The principal sources of air pollution are:
  (a) Domestic heating.
  (b) Motor vehicles.
  (c) Electricity generating stations.
  (d) General industrial sources.

It is clearly of great interest to be able to break down the total damage cost by source of emission. However, in order to do this it is first necessary to establish a relationship between the quantity of pollutant emitted and the measured level of pollution at the point of damage. Many such relationships have been derived, with varying degrees of success. Any calculation must take into account a number of factors as follows:
  (1) Location of source.
  (2) Rate of emission of pollutant.
  (3) Mode of emission (eg height of chimney).
  (4) Mode of dispersal.
    ie (a) wind direction and speed,
    (b) thermal effects,
    (c) atmospheric turbulence,
    (d) variation in topography,
    (e) presence of buildings.
  (5) Absorption from atmosphere.
  (6) Method of measurement.
  (7) Location of instrument.

Each of these factors is subject to a number of random elements which makes accurate prediction unlikely. The only notable attempt to make an allocation of pollution damage costs to sources has been made by Scorer (1957), who used the damage costs estimated by Beaver (1954). Scorer used his experience as a meteorologist to produce weighting factors for sources based on the height and temperature of emission, bulk nature of the source, radiative effects from the pollution cloud, the relationship between source and weather conditions, proximity of the source to the damage site, and the chemical nature of the pollutants. Thus high stacks with hot emission plumes produce effective mixing of discharged gases with a large volume of air so that only a small proportion of the polluting gases can produce effects at ground level within the UK. On the other hand, low-level discharge from domestic heating and motor vehicles give little mixing and most of their discharges are available to produce local damage.

On the basis of the Scorer weighting factors, the following table was derived for use in the PAU study.

Table 10.1
SOURCE WEIGHTING FACTORS FOR POLLUTION DAMAGE COSTS
(after Scorer (1957))

Source of pollution	Weighting factor
Domestic heating	120
Motor vehicles	60
Electricity generation	1
General industrial sources	4

The weighting factor multiplied by the total weight of emission from the source gives the weighting index for that source (normalised so that the sum of the indices for all sources equals unity). The total damage cost can then be divided between the sources in proportion to the weighting indices. The exception is agricultural damage, the cost of which was allocated on the basis of total emissions from each source, without any other factors being introduced.

Control Costs
Although there exists considerable information on the cost of controlling pollutant emissions of different types from various sources, it was found to be very difficult to draw any valid conclusions about total national costs of control. The picture was further complicated by the

fact that in the domestic heating area (currently the most important single source of damage), the cumulative fuel savings resulting from the use of more efficient forms of combustion, coupled with the continued voluntary swing towards the use of low-sulphur fuel oils and natural gas heating systems, meant that well before the end of the century, the control costs would be negative in this area. That is to say the savings gained would far outweigh the extra costs of installing new domestic heating equipment.

Treatment of Uncertainty

Referring back to Fig 10.1, any error in locating the precise value of Y implies an error in the location of O. Since O is the optimum point, an error in its position implies a loss of maximum benefit B, either from failure to reduce pollution level sufficiently, or from an excessive reduction. Thus the expectation of error in O = expectation of loss of benefit. If $f(B)$ defines the probability density function of the maximum benefit B, and $\bar{B}$ is the median value of B, then the expectation of error incurred in taking $\bar{B}$ to be the actual value of B is given by:

$$\int_B |B - \bar{B}| . f(B) . dB$$

where:

$|B - \bar{B}|$ = the absolute value of the error

Conversely, this is also equal to the expected gain in maximum benefit resulting from the complete removal of uncertainty surrounding the location of O. This therefore sets an upper limit to the justifiable spend on investigational work aimed at removing uncertainty. This overall figure can be disaggregated to give the justifiable spend in removing the uncertainty surrounding the various components of the total cost.

Pollution Damage to Agriculture

Because, in the main, major agricultural areas are not interspersed with areas of high industrial density, but are separate from them, the

effects of air pollution on agriculture can be divided into two categories:
 (a) Those produced by pollutants which are emitted by a large number of sources all over the country giving general background levels (eg smoke, sulphur dioxide).
 (b) Those produced by pollutants emitted in well-defined places and which are more local in extent (eg fluoride).

*Sulphur dioxide** can affect life in three ways: as a constituent of the air; as rain; or as acidic mist or dust. It may be in the form of sulphur dioxide itself, or oxidised as sulphate or sulphuric acid. When washed out into the soil by rain it can affect the soil and hence the health of the plants. When oxidised to sulphate (a normal and essential constituent of plant tissue) it can cause delayed blooming, premature senescence, and hence lowering of plant resistance to cold, drought, etc. In the gaseous form, above a certain concentration level (which depends on variety of plant and age of leaf) it can cause leaf damage. Conifers are particularly susceptible to damage by sulphur dioxide (Jones (1972)). Animals suffer from the same respiratory effects as humans. Its effects are not all adverse however; it prevents certain fungal diseases of plants such as black spot disease in roses.

Smoke has several effects: it reduces the supply of light radiation required for photosynthesis: it may block the stomatal pores, hindering or preventing the passage of gaseous oxygen, carbon dioxide or water vapour; and it adheres to the glass of greenhouses and diminishes the amount of light reaching the plants. It may help, on the other hand, to reduce frost incidence.

Fluorine can be emitted in gaseous effluent from brickworks and aluminium smelters as a constituent of a number of compounds. Besides retarding growth in plants, as a result of destruction of tissue, it can, when consumed by animals, lead to softening of bone and teeth.

Oxides of nitrogen are emitted in relatively small quantities compared to sulphur dioxide but could grow as the number of cars on the road increases. They are converted into nitrates and find their way on to

* See Chapter 1, herein.

the land, where to some extent they can be beneficial, acting as artificial fertilisers. It is interesting to note that the quantity of nitrates emitted into the air (most of which are rained on to the land) in the United States, amounts to about 3 million tons per annum, compared to a deliberate addition of nitrate fertilisers of 9 million tons. The oxidising action of nitrogen dioxide or nitric acid washed out by rain could produce direct damage to foliage.

Other pollutants which adversely affect vegetation include ethylene (found in motor vehicle exhausts and in gaseous effluents from certain chemical plants such as oil refineries), ozone and organic sulphur compounds (again from oil refineries).

Effects of Air Pollution on Weather and Climate

Any discussion of the effects of air pollution in the meteorological field must of necessity be rather vague and speculative, firstly, because meteorology is not an exact science and, secondly, because the degree of interaction between pollutants and meteorological phenomena is a matter of some dispute.

In this section it is proposed to make the somewhat arbitrary distinction between (a) weather—which is used here to denote effects which are confined to one region, and (b) climate—which is used to denote effects of a global character. Pollution in these respects must be taken to include, for example, carbon dioxide, thermal pollution, jet trails from aircraft, and power station cooling tower plumes.

Weather effects, actual or alleged, can be divided into those concerning (a) rainfall, (b) sunshine and (c) fog. With reference to rainfall, particles of suspended solid matter in the atmosphere can act as nuclei for the condensation of water droplets, leading to cloud formation or rain. Cloud seeding experiments have shown that precipitation can either be increased, decreased, or unaffected, depending on the movement, temperature and humidity of the surrounding air. As far as the UK is concerned, there is no evidence that pollution significantly affects the amount of rainfall but there are indications that the sulphur content of rainwater is higher in regions of high sulphur dioxide pollution.

The obscuration of the sun by fog and smog caused or intensified by air pollution is a subject which has been investigated in many places, including work by the Meteorological Office in this country. Thus, there have been significant increases (about 50 per cent) in the duration of bright sunshine during the months of November to January at the London Weather Centre, since 1958, compared with the long-term average 1931–60 (Jenkins (1969)). In addition there certainly seems to have been a decrease in the frequency of dense fog in central London since the mid-fifties (Brazell (1968)). It should be borne in mind that the Clean Air Act came into force in 1956.*

Climatic effects can be divided into those arising from increased carbon dioxide emissions and those from increased emissions of particulate matter. There is no doubt that carbon dioxide concentrations in the atmosphere have been increasing this century and it is now about 11 per cent higher than the 1850 level, primarily due to the increased combustion of carboniferous fuels and the drainage of peat bogs, etc, releasing trapped gases. The physical effect of increased levels of carbon dioxide, the so-called greenhouse effect, is to cause global warming. However, the fact is that since about 1940 the average northern hemisphere temperature has shown a definite fall, irregularly at first but more sharply since 1960, which indicates that other factors involved are more influential than the rise in carbon dioxide levels. Thus a 10 per cent change in carbon dioxide level could be counterbalanced by a 3 per cent change in water vapour or a 1 per cent change in mean cloudiness. Two great unknowns are the powers of the oceans and world vegetation to emit or absorb carbon dioxide. However, it is clear that releases from man's activities will become more dominant as time goes on.†
Particulates are being emitted from most industrial and domestic sources at ever-increasing rates, with a doubling time of 10–15 years. It has been

* See Chapter 9, herein.
† See Chapter 1, herein.

calculated (Bryson (1968)) that a reduced atmospheric transparency of 3–4 per cent could decrease the mean global temperature by 0·4° C. It must be borne in mind, however, that man's emissions of particulates are small in quantity compared with those from volcanoes, dust storms, etc, but they are beginning to approach the average values from natural sources. Summarising these counteracting climatic effects, it appears that man's release of pollutants into the atmosphere have so far had little impact on the global climate, but that within the next thirty years or so this impact will be increasingly significant, and possibly dominant, unless some counteracting measures are taken. A further complicating factor is the increase of thermal pollution (at present 0·04 per cent of the solar radiation retained by the earth's surface (Budkyo (1962)) and increasing at 4 per cent per annum, compound). The risk of triggering major swings in global climate within two generations from now is not infinitesimal, and the potential impact these would have on the economic and social conditions in all nations is incalculable.

Conclusions

In view of the fact that a valid assessment of control costs was not made, the results could not be used to determine the optimum pollution levels. However, the estimates of damage costs, broken down by source, were of considerable value in guiding future R & D activity and this was the prime object of the study. An analysis was carried out to determine in which parameters the reduction of uncertainty would have the greatest value. This enabled a ranking of R & D priorities to be made. From the study it was clear that much work remains to be done in establishing the true nature of the damage cost/pollution level relationships, at the relatively low levels of pollution encountered in the environment. A major difficulty in laboratory control work is in measuring damage effects at these low levels over anything like a reasonable period of time. This applies particularly to agricultural damage.

Mention has been made earlier of the distinction between economic and social costs. This distinction appears most forcefully in the assessment of damage to health. Economic costs include items such as loss of production due to illness from respiratory disease, cost of treatment, and in extreme cases, the economic consequences of an early death. However, the 'social costs' of suffering and early death are not included, yet it could be argued that these are the more important of the two, and certainly any government must take these into account when formulating its policies. The question is: how is one to add together social costs and economic costs so that the optimum control solution can be found? No solution to this question is offered. Certainly, a 'value' of injury and life is implied by court awards, by the sums spent on reducing road accidents, and by the sums spent on keeping non-productive members of society (eg sick, old people) alive. The unfortunate conclusion, as far as cost/benefit analysis is concerned, is that these implied social costs are wildly inconsistent, which is not to be unexpected in so emotionally sensitive a field. Here the matter rests, for the time being. It clearly remains a major task for government to find a satisfactory balance between economic and social costs in formulating its policies, whether in pollution control, or elsewhere.

References
BEAVER, SIR HUGH. *Report of the Committee on Air Pollution* (HMSO, 1954).
BRAZELL, J. H. *London Weather* (HMSO, 1968).
BRYSON, R. S. 'All Other Factors Being Constant', *Weatherwise*, **21**, No 2 (1968), 56–61.
BUDKYO, M. I. 'Certain Means of Climate Modification', *Meteorologiia i gidrologiia*, No 2 (1962), 3–8.
JENKINS, I. 'Increases in Averages of Sunshine in Greater London', *Weather*, **24**, No 2 (1969), 52–4.
JONES, G. E. 'An Investigation into the Possible Causes of Poor Growth of Sitka Spruce (*Picea sitchensis*), in Taylor, J. A. (ed), *Research Papers in Forest Meteorology: An Aberystwyth Symposium*, **XII** (1972), 147–56.
SCORER, R. S. 'The Cost in Britain of Air Pollution from Different Types of Source', *J Inst Fuel*, **XXX**, No 194 (1957), 110–15.

CHAPTER 11

P. J. CODLING

Weather and Road Accidents

Introduction
This paper describes research work on the effect of weather on road accidents in Great Britain undertaken at the Transport and Road Research Laboratory. The purpose of the research was (a) to find the association of the various kinds of weather with traffic flow and types of accident, and (b) to estimate numerical factors that can be used to take account of changes of weather for understanding variations in traffic flow and numbers of accidents. In addition, the assessment of the economic consequences of weather was also necessary in justifying the cost of counter-measures.

Various approaches have been used in studying the effects of weather on traffic flow and accidents. This study describes more recent analyses. The different road-weather conditions—rain and wet roads, snow and ice, and fog—are considered separately.

Weather and People's Movements
The amount and density of traffic flow are influenced by changes in weather since some people are less likely to go outdoors if the weather is bad. The pattern of travelling will be affected by those who decide that journeys are not essential, resulting in cancellation or postponement of such journeys, or the resort to a different form of transport. The effect is likely to vary according to the time of day at which the bad weather occurs. For instance, if bad weather occurs early in the day a person's reaction may be to delay the journey in the hope that the weather will improve, whereas if it occurs later in the day he may terminate his journey early. It is also possible that some people may

cancel journeys because they believe the weather is going to be bad, whether or not it actually turns out to be so. On the other hand, there may be those who decide to travel the following day because the weather forecast is good.

Some of these decisions imply that traffic at a particular time is affected by *previous* weather. In the present study, however, traffic and accidents have been correlated with *current* weather conditions and the above considerations have been largely ignored.

Sources of Data

Weather information has been obtained mainly from the Daily Weather Reports, the Monthly and Annual Summary of the Monthly Weather Reports of the Meteorological Office (published by HMSO). Some data have also been obtained directly from the Meteorological Office and Weather Centres.

Traffic data were obtained mainly from automatic traffic counters at road sites. Since January 1956, automatic counters providing daily totals have been in continuous operation at fifty census points distributed according to statistical principles to give a representative sample of flow on the classified road system in Great Britain (Tanner and Scott (1962)). The main purpose of these counts is to provide information about national trends in road usage from one period to another.

Accident data were obtained from the National Accident Coded Sheet (Form Stats 19), which is completed by the police in respect of each personal injury accident reported to them. National road accident statistics are computed from these data. Since 1 January 1969, Stats 19 has specified the weather (rain, snow, fog, other) as well as the road surface condition (wet, dry, icy) at the scene of an accident.

Assessing the Effects of Weather

To assess the effects of weather it is necessary to compare the amount of traffic and numbers of accidents in a given period of bad weather with those which would have been expected to occur in that period if the weather had been fine. In practice this is not as simple as it sounds because of the limited data available, which explains why a variety of approaches have been used in the various investigations covered by this paper.

Weather can operate directly as rain, snow or fog, or indirectly through the state of the road surface being wet, dry or icy. The frequency of road accidents is influenced by it in two ways. It affects the amount and type of traffic on the road, and hence the number of road-users exposed to risk, and it also affects the risks per unit of travel by creating additional hazards such as slippery roads and poor visibility.

Review of Past Work

The main researches into the effects of weather on traffic and accidents, which were carried out by the Transport and Road Research Laboratory in earlier years, are detailed in Tanner, J. C. (1952a, 1952b and 1967), and DSIR (1963). The results show important general trends which may be summarised briefly as follows:

> Wet weather decreased traffic flow but increased the number of accidents and casualties, with the resultant effect of increasing the accident and casualty rates per unit of travel (veh.Km). All kinds of traffic were affected, with the greatest reductions in flow in the case of two-wheeled vehicles.
>
> Snow and ice also reduced traffic appreciably, the greatest reductions again occurring in the numbers of two-wheeled vehicles. For accidents, however, the effect depended on the extent of ice and snow: moderate proportions led to more accidents, while larger amounts led to fewer accidents than expected under dry conditions.
>
> Fog reduced traffic appreciably, with much greater reductions at weekends. Accidents overall increased in number, but one class of injury, namely pedestrian, was reduced in number.

While these effects are quantified in Tanner, J. C. (1952a, 1952b and 1967) and DSIR (1963) their magnitudes are to some extent of only historical interest and do not necessarily apply today. Over the years, changes in vehicle design, environmental factors and legal changes have affected driving habits and conditions, as will be illustrated by some of the more recent work.

Recent Work

Since 1969, the brief references to weather in addition to road surface condition on the national accident report form, have made it possible to examine, for the first time for many years, the incidence of injury accidents occurring on dry, wet or icy roads according to whether it is raining, snowing, foggy or otherwise. Table 11.1 tabulates the data for Great Britain in 1970: the data are similar to those for 1969, neither year being abnormal as far as weather is concerned.

Table 11.1 shows that in 1970, a year with rainfall only slightly above average, 31 per cent of injury accidents occurred on wet roads, and in half of these rain was falling. Icy roads prevailed in 4 per cent of injury accidents, and in one-third of these it was snowing. Fog was reported in just over 1 per cent of injury accidents. Thus, in terms of numbers, the biggest weather problem is associated with rain and wet roads.

The relative severity of injury accidents in different weather conditions is indicated in Table 11.2 by the percentage of injury accidents which were fatal or serious. Results confirm earlier findings of the greater degree of severity on the faster roads and at night, and also show that on average the severity of accidents is much the same on dry or wet roads, in rain or in fog. In snow and on icy roads accidents are on average less severe.

For more detailed analyses of the effects of weather on accidents and traffic, the different conditions will now be considered separately.

Rain and Wet Roads

To assess the effect of rain on traffic flow two analyses were undertaken. In the first, meteorological data were used to identify predominantly

Table 11.1

PERSONAL-INJURY ACCIDENT DATA BY WEATHER FOR GREAT BRITAIN
1970

		Accidents		Casualties	
		Number	Percentage	Number	Percentage
Dry roads					
	Rain	333	0·2	447	0·2
	Snow	20	0·0	30	0·0
	Fog	615	0·4	910	0·4
	Other	173,099	99·4	229,452	99·4
	Total	174,067	100·0	230,839	100·0
Wet roads					
	Rain	40,765	49·2	57,337	48·9
	Snow	1,186	1·4	1,594	1·4
	Fog	2,093	2·5	3,180	2·7
	Other	38,902	46·9	55,140	47·0
	Total	82,946	100·0	117,251	100·0
Icy roads					
	Rain	180	1·7	292	1·9
	Snow	3,590	34·4	5,153	34·2
	Fog	468	4·5	670	4·5
	Other	6,194	59·4	8,935	59·4
	Total	10,432	100·0	15,050	100·0
All roads					
	Rain	41,278	15·4	58,076	16·0
	Snow	4,796	1·8	6,777	1·9
	Fog	3,176	1·2	4,760	1·3
	Other	218,195	81·6	293,527	80·8
	Total	267,445	100·0	363,140	100·0
Road conditions					
	Dry	174,067	65·1	230,839	63·6
	Wet	82,946	31·0	117,251	32·3
	Icy	10,432	3·9	15,050	4·1
	Total	267,445	100·0	363,140	100·0

Table 11.2
SEVERITY OF ACCIDENTS IN RELATION TO WEATHER AND ROAD SURFACE CONDITIONS FOR GREAT BRITAIN, 1970

Average severity $\left(\dfrac{Fatal + Serious}{Total} \text{ accidents} \times 100\right)$

		Daylight		Darkness	
		Speed limit		Speed limit	
		40mph or less	Greater than 40mph	40mph or less	Greater than 40mph
Weather					
	Rain	24	40	32	46
	Snow	22	36	28	35
	Fog	26	41	33	45
	Other	25	41	33	48
	Total	25	41	33	47
Road surface conditions					
	Dry	25	42	33	49
	Wet	25	40	33	46
	Icy	22	38	28	38
	Total	25	41	33	47

rainy periods and corresponding dry periods which would be used for comparison. Rainfall records giving duration of rain for 5 weather stations in London for 1969 were studied and the rainiest days common to all 5 stations were noted. There were 7 of these, all weekdays (Monday to Thursday), during which rain fell intermittently throughout the day. Six traffic census sites were within the area enclosed by the 5 meteorological stations. Traffic data recorded at these sites on rainy days were compared with mean counts on 2 corresponding days which were dry. These 2 dry days were the ones exactly a week before and after the rainy day. The mean count was regarded as a reasonable estimate of the traffic flow that would have been recorded if the rainy days had been dry instead of wet. On average over the 7 days, 5 of which were in winter, the flow of vehicles was 98 per cent of the expected flow for a fine day.

In the second analysis dealing with traffic flow, the national accident statistics were used to identify 'rainy' and 'dry' days, so that comparison of traffic recorded at the 50-point census sites referred to earlier (Tanner and Scott (1962)) could be made for these days. A 'rainy' day was defined as one on which at least 50 per cent of casualties were reported in rain. A 'dry' day was one on which less than 10 per cent of casualties occurred in rain. Rainy days without two corresponding dry days exactly a week or fortnight before and after the rainy one were rejected, leaving 16 rainy days for study during the 2 years 1969–70. It was assumed that rainy days defined in this way were wet enough to affect people's travel. Table 11.3 shows that overall the traffic flow was reduced on rainy days. The average reduction on weekdays (Monday–Friday) was just over 1 per cent, but on Sundays the reduction was much greater. Variations on different classes of road were not significant.

Meteorological data have also been used to assess the effect of rain and road wetness on numbers of accidents. Rainfall records for Newcastle, Manchester, Southampton and Kensington Palace were examined for at least one of the two years 1968 and 1969. Records of duration of rainfall were used to identify continuously rainy periods. Amounts of rain, coupled with its duration, were used to identify periods when roads were wet without rain: these were defined as hours for which 'trace' of rain (less than 0·05mm) was recorded but no measurable duration of rain, provided these hours lay between hours of continuous rain (it was assumed that roads would then be wet for the whole of the periods noted but the rainfall negligible). Accident data for each rainy period were then compared in turn with the mean of two corresponding dry periods, and for each wet-road period and the mean of two corresponding dry periods. The dry periods were taken to be exactly a week before and after the rainy or wet one, and their means regarded as a fair estimate of the number of accidents which would have occurred in the rainy period if it had been dry instead of

Table 11.3

TRAFFIC FLOW CHANGES ON THE RAINIEST DAYS DURING 1969–70 IN GREAT BRITAIN

(TRUNK, PRINCIPAL AND CLASSIFIED ROADS ONLY)

Day of week	Month	Motor-vehicle-kilometres nationally	Average percentage change in Daily flow of vehicles on following classes of road			
			Trunk (10 sites)	I (16 sites)	II (9 sites)	III (15 sites)
Sunday	January	−17·8	−13·4 (8)	−24·2 (13)	−27·6 (9)	−27·4 (13)
	September	− 7·6	− 6·5 (7)	−11·9 (12)	−11·2 (8)	− 9·3 (15)
	November	− 7·2	− 7·8 (8)	−11·0 (12)	−22·3 (6)	−17·2 (13)
Monday	January	− 0·3	− 0·7 (8)	+ 0·8 (10)	+ 4·7 (6)	+ 4·8 (11)
	February	+ 0·6	− 0·2 (6)	+ 1·2 (11)	+ 6·0 (8)	+ 5·1 (11)
	June	− 1·9	+ 0·2 (7)	+ 1·1 (12)	− 6·6 (7)	−13·3 (14)
	July	+ 0·9	+ 4·9 (8)	− 2·8 (10)	− 3·6 (9)	− 8·5 (14)
Tuesday	May	− 1·6	+ 0·1 (7)	− 4·8 (14)	− 7·5 (7)	− 8·3 (11)
	July	− 0·4	− 1·0 (8)	+ 0·1 (10)	− 2·6 (9)	+ 2·0 (14)
	October	− 0·6	+ 0·2 (9)	− 1·4 (10)	+ 0·4 (6)	− 0·7 (12)
	November	− 2·8	− 6·9 (6)	− 2·5 (12)	− 8·2 (7)	+ 0·1 (14)
	November	− 2·7	− 1·9 (7)	− 1·8 (9)	− 4·7 (7)	− 1·8 (11)
Wednesday	March	− 1·0	− 1·6 (7)	− 0·2 (9)	− 4·0 (8)	− 5·5 (13)
	August	− 2·6	+ 4·1 (7)	− 4·9 (14)	− 5·4 (9)	−14·2 (14)
Friday	July	− 3·2	− 2·3 (7)	− 3·0 (12)	− 8·0 (9)	− 8·3 (15)
Saturday	February	+ 1·6	− 0·1 (8)	− 0·4 (10)	+ 2·5 (8)	+ 3·6 (12)
Means (Monday–Friday)						
Winter (Jan–March + Oct–Dec)		− 1·1	− 1·8	− 0·6	− 1·0	+ 0·3
Summer (April–Sept)		− 1·5	+ 1·0	− 2·4	− 5·6	− 8·4
Grand		− 1·3	− 0·4	− 1·5	− 3·3	− 4·0

Figures in brackets denote number of 50-point traffic census sites from which data were wet. Table 11.4 summarises the results of the analyses for injury accidents (and casualties) in daylight and darkness, in winter (January–March, October–December) and in summer (April–September).

Table 11.4

EFFECT OF RAIN (CONTINUOUS) AND WET ROADS (NOT RAINING) ON PERSONAL-INJURY ACCIDENTS AT SELECTED PLACES IN 1968–9

	Daylight		Darkness	
	Jan–Mar and Oct–Dec	April–Sept	Jan–Mar and Oct–Dec	April–Sept
Accidents				
Numbers in rain	217	234	231	76
Numbers in dry	149	188	120	43
Increase in rain (per cent)	46*	24*	92*	77*
Numbers on wet roads	142	166	93	(20)
Numbers on dry roads	110	102	62	(8)
Increase on wet roads (per cent)	29*	63*	50*	–
Casualties				
Numbers in rain	249	291	305	114
Numbers in dry	168	218	162	55
Increase in rain (per cent)	48*	33*	88*	107*
Numbers on wet roads	161	194	115	(28)
Numbers on dry roads	124	126	79	(10)
Increase on wet roads (per cent)	30*	54*	46*	–

* Statistically significant change at the 5 per cent level.

The accidents and casualties were more numerous both in rain and on wet-roads-without-rain, than in dry weather. The percentage increase in injury accidents in each case was greater in darkness than in daylight, though no comparison for wet roads in summer was possible owing to the small numbers of accidents reported in the dark under these conditions. The increase in accidents during rain was much greater in winter than in summer, both in daylight (46 per cent compared with 24 per cent) and in darkness (92 per cent compared with 77 per cent). In contrast, the increase in accidents on wet-roads-without-rain (during daylight) was less in winter than in summer (29 per cent compared with 63 per cent). The reasons for these seasonal differences are not immediately apparent, but they could be associated with the different effect on accidents of impaired visibility in rain and increased slipperiness of wet roads in summer. On the one hand, since wet roads are more slippery in summer than in winter, a greater increase in accidents would be expected in the summer: on the other, rain falling in the darker days of winter might impair visibility to a greater extent than in summer, with a consequent greater increase in winter accidents. Nor is it clear why in summer the effect of rain is so much less than that of wet-roads-without-rain. However, another factor which may be relevant is that pedestrians' habits appear to be different summer and winter, there being a relatively smaller proportion of accidents involving pedestrians in rain in the summer.

No direct analysis of the effect of rain and wet roads on accident rates has been carried out, but it is worth noting that since their effect on traffic flow is generally so small, the changes in accident rates will be similar to the changes in numbers of accidents.

Snow and Ice

Analyses of the national accident statistics to identify predominantly snowy days have been made parallel to the analyses for rainy and wet days. 'Snowy' days were defined as those on which at least 40 per cent of casualties were reported in snow (analyses showed these to be the most widespread snowy days nationally). Snowy days without 2 corresponding dry days exactly a week or fortnight before and after, were rejected, leaving 6 snowy weekdays for the 2 years 1969–70 for which comparisons could be made. The traffic and accident data for these 6 days were compared with the means of the corresponding 2 dry days. Thus Table 11.5 shows that on the snowy days traffic was reduced by

12 to 26 per cent, casualties were increased by between 4 and 52 per cent, with the overall effect of increasing casualty rates by 13 to 78 per cent: on average the increase in casualties was 24 per cent and in casualty rates 47 per cent. From comparison with the percentage of casualties occurring when snow was falling and the percentage occurring on icy roads, it would appear that the increases in casualty rates are associated more with the icy road condition than with the falling snow.

Table 11.5

CHANGES IN TRAFFIC FLOW AND CASUALTIES IN GREAT BRITAIN ON THE SIX SNOWIEST DAYS DURING 1969–70

Traffic: Decrease in motor veh-km travelled (%)	Casualties per day				Increase in casualties on snowy days (%)	Increase in casualties per motor veh-km (%)
		On snowy days		On dry days total no		
	Total no	Snow falling (%)	Icy roads (%)			
15	1,571	59	74	1,036	52*	78*
18	1,215	65	72	894	36*	60*
26	857	54	71	781	10*	50*
13	1,080	62	58	948	14*	29*
12	870	42	56	744	17*	30*
15	781	45	46	749	4	13*

* Statistically significant change at the 5 per cent level.

Fog

It has already been seen that accidents in fog account for only a small proportion of the total of injury accidents (Table 11.1), and that overall the severity of accidents in fog is no greater than in clear weather (Table 11.2). However, in view of the current interest in the effects of fog on accidents, especially on motorways, it is important to examine this problem more fully.

One investigation, which has already been reported (Codling (1971)), concerned the occurrence of thick fog (visibility less than 200m) in Great Britain for the period 1958–67 and its effect on traffic flow and accidents. Fog of this degree was found to be relatively infrequent, patchy, rarely widespread and of short duration. The frequency of fog inland was falling over the years probably because of the Clean Air Act, 1956. Only 4 whole days of widespread thick fog (ie an extensive area of continuous thick fog) were found to have occurred since the smog of 3–7 December 1962. On these occasions (two Tuesdays and two Wednesdays) traffic flow was reduced by about 20 per cent. The effect, although by no means consistent at the different places, was similar on all classes of road. Variable changes in the total numbers of accidents and casualties occurred in different fogbound areas. However, the data were too scanty to permit detailed analysis in an attempt to show different effects in different fogs. The various accident and casualty data were therefore summed to give larger samples for statistical testing. It was found that in thick fog the numbers of fatal and serious accidents and casualties were reduced but slight accidents and casualties increased significantly: the increase in total injury accidents was 16 per cent. Accidents in darkness and those involving pedestrians were reduced significantly whereas those involving more than two vehicles were increased significantly. Overall there was no change in the fatal and serious accident rate per unit of traffic whereas the slight injury and total accident rates increased significantly by about 70 and 50 per cent, respectively.

Examination of the accidents on motorways in 1969 and 1970 (see Table 11.6) shows that on average about 60 per cent of injury accidents in fog on motorways involved more than two vehicles, compared with about 10 per cent in clear weather. Also, accidents on motorways in those years were much more severe in fog than in clear weather conditions: the proportion of accidents which were fatal or serious was about 60 per cent in fog compared with about 40 per cent in clear conditions. In terms of numbers of accidents, those occurring in fog are few, but they are more likely to be multiple accidents, and fog on motorways occasionally leads to dramatic and costly pile-ups; the economic aspects of these will be considered later.

Table 11.6

PERSONAL-INJURY ACCIDENTS AND REPORTED WEATHER CONDITIONS ON MOTORWAYS, 1969–70

	Rain		Snow		Fog		Other	
	1969	*1970*	*1969*	*1970*	*1969*	*1970*	*1969*	*1970*
Number of accidents reported involving								
1 motor vehicle	86	107	21	19	14	10	571	697
2 motor vehicles	105	95	27	24	29	8	534	625
3 motor vehicles	35	24	4	1	17	16	104	114
4 motor vehicles	12	14	–	4	11	8	18	36
5 motor vehicles	4	2	–	1	3	3	10	15
6 motor vehicles	–	2	–	–	3	1	6	6
7 motor vehicles	2	2	–	1	1	–	1	1
8 motor vehicles	1	–	–	–	4	3	1	1
9 or more motor vehicles	–	–	–	–	11	6	2	3
Total accidents	245	246	52	50	93	55	1,247	1,498
Accidents involving more than 2 motor vehicles:								
Number	54	44	4	7	50	37	142	176
Percentage	22·0	17·9	7·7	14·0	53·7	67·3	11·4	11·8

218   CLIMATIC RESOURCES AND ECONOMIC ACTIVITY

Slippery Roads
A major problem associated with bad weather is that roads become more slippery when wet or icy and therefore the risk of accidents involving skidding is increased. The importance of skidding as a factor in road accidents has been recognised over many years and improvements are continually being made in anti-skid properties of road surfaces, tyres and braking systems.

It is not proposed to consider skidding as a factor in accidents in detail in this paper though it is useful to record the latest summary of data available. The numbers and percentages of accidents reported as involving skidding in rain, snow, fog and other weather, and on dry, wet and icy roads in Great Britain, 1970 are given in Table 11.7. These data are also similar to those for 1969. Neither year was abnormal as far as weather was concerned.

Table 11.7
SKIDDING REPORTED IN PERSONAL-INJURY ACCIDENTS, 1970

		Number involving skidding	*Total number*	*Percentage involving skidding*
Weather				
	Rain	11,031	41,360	26·7
	Snow	2,754	4,815	57·2
	Fog	1,024	3,184	32·2
	Other	36,467	218,528	16·7
	Total	51,276	267,887	19·1
Road conditions				
	Dry	22,444	174,335	12·9
	Wet	21,551	83,097	25·9
	Icy	7,280	10,449	69·7
	Total	51,275	267,881	19·1

The most recent detailed investigations into skidding in accidents show among other things the well-established seasonal pattern of skidding: under both dry and wet road conditions the percentages of accidents involving skidding were higher in the summer months than during the winter months—see Sabey and Storie (1968) and Road Research Laboratory (1970).

Cost of Accidents

While analyses covered by this paper are of personal-injury accidents only, in calculating the costs of accidents associated with inclement weather, it is possible to include the cost of damage accidents as well. The current method of costing accidents (Dawson (1971)) is such that the estimate of the average cost of an injury accident makes allowance for the many damage-only accidents (approximately 6 times the number of injury accidents), which are generally neither reported nor costed. The value of this estimate is £1,600 in 1970 terms and hence the total cost of all accidents in any weather condition can be obtained by multiplying the number of injury accidents reported in that condition by £1,600.

As shown in this paper, bad weather has only a small effect on the severity of accidents, so the same average costs can be applied to each weather condition except in the case of accidents on motorways in fog, which will be considered separately.

The numbers of injury accidents in inclement weather in 1970 can be grouped conveniently for costing purposes as follows:

81,000 in wet weather (rain or wet roads), excluding fog;

10,000 in snow or ice, excluding fog;

3,000 in fog;

making a total of 94,000.

The total costs of accidents in these weather conditions were estimated as: £130 million in wet weather; £16 million in snow or ice; £5 million in fog. For comparison, 173,000 accidents occurred in clear, dry conditions in 1970, at an estimated cost of £277 million. The estimates include costs of medical treatment, damage to vehicles and property, administrative costs of police and insurance, loss of output, suffering and bereavement, but do not include cost of delays to vehicles not directly involved.

It has been shown that the increases in accidents in bad weather over the expected numbers average about 50 per cent on wet roads, 25 per cent on snow or ice, and 16 per cent in fog. The costs of these excesses were approximately £46 million, made up of: £42 million in wet weather; £3 million in snow or ice; and less than £1 million in fog. The last figure must be qualified by pointing out that the cumulative cost of the occasional multiple pile-up on a motorway, in which several fatalities occur and up to two hundred vehicles may be involved, is of the order of several hundred thousand pounds. Even so, this does not affect appreciably the total increased cost of accidents attributable to bad weather, which amounted to £46 million in 1970.

Conclusions

A number of studies have been made over many years of the effect of different weather conditions on traffic flow and accident rates. It is not easy to compare the results (a) partly because some refer only to hourly, daily or monthly frequencies, (b) partly because of differences in the methods of measuring weather and (c) partly because of differences in places and periods studied. The various investigations, on the whole, agree that bad weather (rain and wet roads, snow or ice, and fog) reduces traffic flow and increases accident rates, but the magnitudes of the effects vary considerably depending on the basic data used.

Although it was considered useful to make reference to research conducted in earlier years, less significance should be attached to the results, when considering current problems, because of changes in vehicle design, legislation and environmental factors which over the years have affected drivers' habits and conditions. This is particularly noticeable when considering the effect of rain and wet roads on traffic. For example, in the 1950 analysis (Tanner, J. C. (1952b)) the reduction in traffic flow due to rain averaged 6 per cent for weekdays, whereas the analysis in 1969–70 showed a reduction averaging only a little more than 1 per cent. The recent analyses reported in this paper may also be considered more reliable because they are on the whole more direct and more comprehensive. This is largely due to the availability of the additional weather information on the National Police Accident Report Form.

In terms of numbers of accidents the greatest weather problems are associated with rain and wet roads. In 1970, 31 per cent of all injury accidents occurred on wet roads, nearly half of them when rain was falling. From the recent analyses herein, it is estimated that the increase in accidents in rain averages 52 per cent over the whole year, and on wet roads, when it is not actually raining, 50 per cent. The increases are greater in darkness than in daylight. A marked difference in the effect of rain and wet-roads-without-rain occurs between summer and winter. In rain the increase in accidents is higher in winter than in summer: on wet-roads-without-rain the reverse is true. An explanation is offered in terms of relative slipperiness of wet roads in summer and winter, impairment of visibility in rain, and changes in pedestrian habits. The severity of accidents in rain and on wet roads is similar to that in clear, dry weather.

By comparison, although snow, ice, and fog can disrupt the nation's traffic, the numbers of days affected each year are relatively few and the proportions of accidents occurring in these conditions are small. In 1970, only about 4 per cent of all injury accidents occurred on 'icy' roads, and snow was falling in one-third of these. On average, casualties were increased by 25 per cent on snowy days, though the casualty rates per unit of traffic were nearly doubled. Severity of accidents on snow and ice is less than that on dry roads.

Fog was reported in just over 1 per cent of accidents in 1970, which was not an unusual year as far as incidence of fog is concerned. For injury accidents as a whole, the increase in accidents in fog has been estimated at 16 per cent, though as for snow and ice, since the traffic was considerably reduced, the accident rate was about 50 per cent higher than in clear weather. The effect varied with the type of accident.

Accidents in darkness and those involving pedestrians were significantly fewer in fog, whereas those involving more than two vehicles were significantly increased, and accidents were relatively more frequent on fast roads. The greater risk of multiple accidents on motorways occasionally led to a dramatic and costly pile-up.

It is estimated that the average increase in costs of accidents due to bad weather conditions amounts annually to about £42 million in wet weather, £3 million in snow or ice, £1 million in fog (excluding the occasional multiple accident on motorways) making a total of £46 million.

Acknowledgement
This paper is contributed by permission of the Director of the Transport and Road Research Laboratory. Crown Copyright 1972. Reproduced by permission of Her Majesty's Stationery Office.

References
CODLING, P. J. *Thick Fog and Its Effect on Traffic Flow and Accidents*, Department of the Environment, Road Research Laboratory Report No LR 397 (Crowthorne, 1971).
DAWSON, R. F. F. *Current Costs of Road Accidents in Great Britain*, Department of the Environment, Road Research Laboratory Report No LR 396 (Crowthorne, 1971).
DEPARTMENT OF SCIENTIFIC AND INDUSTRIAL RESEARCH, Road Research Laboratory. *Research on Road Safety* (HMSO, 1963).
ROAD RESEARCH LABORATORY. *Skidding in Accidents in 1968*, Road Research Laboratory Leaflet LF 167 (Crowthorne, 1970).
SABEY, BARBARA E. and STORIE, VALERIE J. *Skidding in Personal-Injury Accidents in Great Britain in 1965 and 1966*, Ministry of Transport, Road Research Laboratory Report No LR 173 (Crowthorne, 1968).
TANNER, J. C. 'Effect of Weather on Traffic Flow', *Nature*, **169** (4290) (1952a), 107.
——. 'Weather and Road Traffic Flow', *Weather*, **7** (9) (1952b), 270–5.
——. *Casualty Rates at Easter, Whitsun and August Public Holidays*, Ministry of Transport, Road Research Laboratory Report No LR 74 (Crowthorne, 1967).
TANNER, J. C. and SCOTT, J. R. *50-point Traffic Census—The First Five Years*, Department of Scientific and Industrial Research, Road Research Technical Paper No 63 (HMSO, 1962).

CHAPTER 12   S. R. JOHNSON
                and J. D. McQUIGG

Some Useful Approaches to the Measurement of Economic Relationships which Include Climatic Variables

Introduction
Studies of effects of weather and climatic variables on economic activities have reached an interesting crossroads. The campaign to convince social and physical scientists of the importance of climatic conditions on other than the most obvious of economic and related physical activities has been effectively waged. That is, it is commonly accepted that climatic conditions have an important impact, however subtle in nature, on a majority of the economic activities in which we engage (Maunder (1970)). Having established this, the next step in the progression of scientific inquiry in this area would seem to be the development of measures of these climatic effects which are sufficiently precise to permit the assessment of costs and benefits of the various existing and potential programmes designed to improve outcomes of weather-related activities. These results would of course facilitate more enlightened decisions as to the funding and operation of the numerous private and governmental programmes, directed towards the collection and dissemination of weather information, and more rational consideration of consequences of advertent and inadvertent forms of weather modification.

It is in this second phase of our inquiry that results have been particularly slow in forthcoming. Reasons for the difficulties experienced in developing such information are apparent. The development of experimental data to support the estimation of the impact of climatic conditions on economic activities, with enough precision to be of use in decision-making situations, is an extremely expensive proposition. In fact, except for some rather specific instances, we find that the expense of generating such information has proven prohibitive. As a consequence, the effects of climatic variables have been explored within the context of alternative types of secondary data.

These investigations of effects of climatic variables in economic activities based upon secondary data, however, have met with only limited success. Two of the more important reasons for the lack of success in these endeavours are: (a) the information content of the secondary data and (b) the compatibility of estimated results with the analytical structures of modern decision-making models. In the sections to follow we suggest two estimation methods which can be of use in isolating useful relationships from available data on climatic and economic variables. Both methods are refinements of the technique of regression analysis and thus, can be readily adapted to the current methods of estimating economic relationships which include climatic variables.

Principal Components and Regression Analysis

Data limitations, both in terms of quality and quantity, frequently make it difficult to determine the selective effects of climatic variables in relationships involving economic behaviour and in physical relations which give structure to problems concerning the economic effects of weather conditions. The natural inclination in such situations is to introduce additional information into the estimation procedure. Such information would normally come from results of other empirical investigations, subjective considerations or from generally accepted theories as to the fundamental characteristics of the processes being studied. The exploratory nature of investigations of effects of climatic variables on economic activities, however, tends to make supplemental information of the above types either unavailable or suspect.

One method which can be used to reduce information requirements in least-squares estimation models is based on the use of principal components (Johnson, Reimer and Rothrock (1972)). Since weather variables such as temperature and rainfall for successive periods are frequently highly interrelated, their observed variation is likely to be explained by a limited number of principal components. When this situation prevails, principal components which account for little of the observed variation in the climatic variables, can be eliminated. The elimination of these linear relationships between climatic variables reduces the number of parameters to be estimated. Derived parameter-estimates for all of the climatic variables in the economic and technical relationships can then be deduced from the expressions estimated with the retained set of principal components.

The estimation procedure based upon the use of principal components involves some computations in addition to those employed in standard regression analysis but remains in principle quite simple. To illustrate the procedure let y and u be t-component random vectors related by the structural equation:

$$y = Xb + u \qquad \text{Equation 1}$$

where:

b is a k-component vector defining the linear equation for all values of the t by k matrix of variables X. It is assumed that the dimension of $X (= x_1, \ldots, x_k)$ is $k < t$, the rank of X is k, and that X is nonstochastic. Conditions on the disturbance term are $E(u) = 0$ and $E(uu') = \sigma^2 I$, I being an identity matrix of dimension t and $\sigma^2 \neq 0$ being finite. For purposes of exposition, it is convenient to partition X into two parts $X = [X_1 \vdots X_2]$. $X_1 (= x_1, \ldots, x_{p-1})$ is a set of observations on non-weather variables in Equation 1. $X_2 (= x_p, \ldots, x_k)$ is the set of observations on weather variables.

P

The jth principal component of the sample observations X_2 is the normalised linear combination:

$$z_j = X_2 a_j \qquad \text{Equation 2}$$

The elements of a_j are given by the characteristic vector corresponding to λ_j, the jth characteristic root of $S_2 = X_2'X_2/t-1$. Variances for the principal components are estimated in accordance with usual procedures (Anderson (1958)). For example, the variance of the jth component is:

$$z_j'z_j/t-1 = a_j'X_2'X_2 a_j/t-1 = a_j'S_2 a_j \qquad \text{Equation 3}$$

Since the a_j are characteristic vectors of S_2, it follows that $a_j'(S_2 - \lambda_j I)a_j = 0$. Hence, on suppressing the identity matrix the variance of z_j can be alternatively defined to be equal to the value of the characteristic root λ_j. Thus:

$$\lambda_j / \sum_{j=1}^{k} \lambda_j$$

is the proportion of the variance of S_2 explained by the jth component.

The estimation method based upon principal components is a three-step procedure. First, principal components are computed for X_2. The components are then partitioned according to the amount of cumulative variation in X_2 to be explained.* Call the retained set of principal components Z_1 ($Z = [Z_1 \vdots Z_2]$ being the complete set of components). Secondly, least-squares estimators on the model $y = X_1 b_1 + A_1 d_1 + v$ are calculated. Lastly, the estimators for the climatic variables, $\tilde{b}_2$, are calculated by partially differentiating the estimate of y from phase two, $\hat{y}$, with respect to x_j, ie, $\tilde{b}_j = \delta(\hat{y})/x_j$, ($j = p, p+1, \ldots k$). The derivative is a function of a function so that $\tilde{b}_2$ and $A_1 \hat{d}_1$, where A_1 is

---

* For a discussion of statistically (mean square error) based procedures for determining the omitted components see Johnson, Reimer and Rothrock (1972).

the matrix of vectors a_j defining Z_1 and $\hat{d}_1$ is the least-squares estimator of d_1 from phase two. Variances for the estimators, $\tilde{b}$, are derived by a similar transformation on the variance-covariance matrix for the least-squares estimators from phase two.

An Example Application of the Principal Components Regression Method

Economic rent and/or land price is a surrogate which reflects a number of economic and non-economic factors (Johnson and Haigh (1970)). In estimating the influence of various sets of factors on agricultural land prices, some measure of soil productivity is usually included. Land productivity, however, depends not only on the inherent capacity of the soil but also on the climate. Data from a sample of counties in Arkansas, Illinois, Iowa, Kansas, Missouri, Nebraska, Oklahoma, and South Dakota were employed in estimating the importance of climatic variables in average county land prices. Non-climatic variables in the relationship were used to reflect differences in soil productivity, taxes, governmental subsidies, and non-agricultural demand for land. Climatic variables were monthly temperature and precipitation normals. The resulting linear equation relating average land prices to the climatic and non-climatic variables included 29 variables with the addition of a constant term. The principal components procedure was applied because of the near singularity of the data set on the climatic variables.

Vectors defining the first two sets of principal components for the 12 monthly precipitation and temperature normals are given in Table 12.1. Proportions of the observed variation in the monthly precipitation normals explained by the principal components are 0·64 for the first component and 0·17 for the second component. For the temperature normals, the first component explained 96 per cent of the observed variation and the second component added 3 per cent to the total.

Least-squares estimates of the equations including the two principal components for temperature and precipitation are included in Table 12.2. Estimates from two census years are reported. With the exception

Table 12.1

MONTHLY COEFFICIENTS FOR EIGENVECTORS FROM MONTHLY TEMPERATURE AND PRECIPITATION NORMALS (USA)

Eigen-vector	Jan	Feb	March	April	May	June	July	August	Sept	Oct	Nov	Dec	Portion of total variance explained
Prec$_1$*	0·157	0·315	0·335	0·343	0·295	0·180	0·274	0·188	0·293	0·330	0·342	0·324	0·64
Prec$_2$†	0·111	0·330	0·223	0·124	−0·093	−0·534	−0·234	−0·458	−0·370	−0·036	0·177	0·292	0·17
Temp$_1$‡	0·286	0·288	0·292	0·292	0·288	0·288	0·280	0·283	0·293	0·292	0·292	0·289	0·96
Temp$_2$§	−0·413	−0·331	−0·203	−0·025	0·122	0·285	0·498	0·403	0·159	0·050	−0·202	−0·321	0·03

* Coefficients of the first principal component for the 12 monthly precipitation normals.
† Coefficients of the second principal component for the 12 monthly precipitation normals.
‡ Coefficients of the first principal component for the 12 monthly temperature normals.
§ Coefficients of the second principal component for the 12 monthly temperature normals.

of the constant term, the parameter estimates in Table 12.2 are statistically significant at high probabilistic levels of rejection. Values of parameter means for government payments are 9·45 in 1964 and 6·38 in 1959. The 9·45 estimate for 1964 indicates that the payments are capitalised into the land value at a discount rate of approximately 10 per cent. The lower coefficient on payments for 1959 simply reflects the lower interest rates at the time. Property taxes appear to be capitalised into the value of land at about the same rate as subsidies. Parameter estimates for population and soil capability are not as easily interpretable but have proper signs and are consistent between years.

Effects of precipitation and temperature normals are calculated as previously described. Because of the similarity between the 1964 and 1959 regressions in Table 12.2, the derived least-squares estimators are calculated only for 1964. Mean estimates calculated using the components in Table 12.1 and the mean parameter estimates in Table 12.2 are presented in Table 12.3. They are presented both as directly calculated and as percentages of average land price. In the latter case, units of measure are divided away so that measurement problems associated with the use of census data are reduced. With two minor exceptions, the parameter estimates indicate that adding to precipitation normals can increase average land price. In addition, the relative importance of monthly precipitation normals corresponds to the growing season for crops in the area. Temperature normals are less important and less easily interpreted. However, increases in these normals appear to have positive effects on land prices at the end of the growing season and have negative effects during typically hot growing season months.

The land price problem is but one of a number of situations in which the principal components regression procedure has been applied. As further examples: (1) we have applied principal components regression in estimating production functions for maize, using experimental plot data from many locations in the 'corn belt' of the US together with observed temperature and rainfall values (Benson and Johnson (1970));

Table 12.2

AVERAGE LAND PRICES FOR COUNTIES RELATED TO PROPERTY TAX, GOVERNMENT PAYMENTS, SOIL CAPABILITY, POPULATION DENSITY, AND THE PRINCIPAL COMPONENTS ON CLIMATIC VARIABLES (USA)

Year	Climatic variables				Non-climatic variables				Constant term	Coefficient of determination
	$Temp_1$	$Temp_2$	$Precip_1$	$Precip_2$	Soil capability index	Population density	Property tax per acre	Gov't payments per acre		
1964	−4·23 (0·44)	−10·40 (1·61)	10·90 (1·78)	−13·28 (3·15)	60·40 (4·57)	3·32 (0·40)	−11·70 (3·00)	9·45 (1·48)	108·87 (67·25)	0·77
1959	−4·06 (0·40)	−9·02 (1·45)	9·28 (1·59)	−16·47 (2·82)	57·30 (4·09)	2·27 (0·36)	−3·82 (2·65)	6·38 (1·33)	97·22 (60·21)	0·79

Note: Terms in parentheses are estimated standard errors.

(2) we have used the eigenvalue approach to estimate the amount of diversity that can be expected within a large interconnected electric power system during periods of critical weather (McQuigg et al (1972)); (3) we are experimenting with principal components regression in development of more precise weather-load models for electric power systems.

Table 12.3
INFLUENCES OF CHANGES IN MONTHLY TEMPERATURE AND PRECIPITATION NORMALS ON AVERAGE PRICES OF AGRICULTURAL LAND IN COUNTIES (USA)

Month	Directly calculated		Percentage of average land price	
	Temperature	Precipitation	Temperature	Precipitation
January	3·10	0·23	1·7	0·1
February	2·23	−0·96	1·2	−0·5
March	0·88	0·68	0·5	0·4
April	−0·97	2·09	−0·5	1·1
May	−2·47	4·46	−1·3	2·4
June	−4·17	9·06	−2·3	4·9
July	−6·36	6·10	−3·5	3·3
August	−5·38	8·14	−2·9	4·4
September	−2·88	8·11	−1·6	4·4
October	−1·75	4·08	−1·0	2·2
November	0·87	1·38	0·5	0·7
December	2·13	−0·35	1·2	−0·2

A rational model in this last application has to involve more than one meteorological variable, eg, dry bulb temperature for successive three-day periods, wind, and cloud cover. These variables are highly intercorrelated. Applications of the method give reductions in the variance of the regression coefficients, compared to the classical least squares regression model. This has important implications for both short-term dispatchers and for long-term planners in this vital industry. In each of these cases the results have shown to be an improvement over those obtained with standard regression analysis.

Linear Probability Models and Regression Methods

Probabilistic statements concerning the feasibility of conducting operational activities are common input for various types of management and economic decision models (Hiller and Lieberman (1967)). Moreover, it is frequently the case that these probabilities are conditioned by the selected types of observable climatic events. The feasibility of carrying out many construction activities is commonly dependent upon precipitation, temperature and wind conditions. Agricultural crop and livestock production activities are dependent upon the same sorts of climatic conditions; susceptibility of human and animal populations to diseases are conditioned by climatic events and the like.

Until recently, these conditional probabilistic statements, which are ultimately of consequence in decision-making and thus concern the economic consequences of the climate, seemed to be introduced through rules of thumb and other mechanisms based upon experience but not objectively derived. The linear probability model to be discussed in this section represents a method for the objective determination of such weather probabilities. Since the 'sample' data for such estimations can be passively generated, ie, taken from operating records of companies, individuals or public institutions, this method of estimation may represent a useful way to obtain more precise measurements of weather effects upon economic activities while using only available data and computational resources.

The form of the linear probability model is the same as that given in *Equation 1*. As in the case of the discussion of the principal components regression methods, the conditioning variables may be climatic or non-climatic in nature. Hence, the partitioning discussed in that connection may be preserved but will not be of consequence in the current discourse.* Given the specification in *Equation 1*, the probabilities, ($\hat{y}$), are to be estimated from samples of concomitant observa-

---

* In the event of severe problems of collinearity, however, the principal components method could be used in conjunction with the linear probability models discussed in this section.

tions on the set of independent variables, $x_1, x_2, \ldots, x_k$, and an artificial variable, y, taking the value zero or one depending upon whether or not the event of interest did or did not occur. The estimation problems presented by such models are two-fold. First, the errors are not distributed as usually assumed. Secondly, the estimation must be carried out in such a way as to guarantee that the estimated values of the probabilities fall in the unit interval. The former problem can be handled by using generalised least squares methods of estimation (Goldberger (1964)). The latter suggests non-linear estimation procedures called probit or normit analysis and logit analysis. Although these estimation methods require specialised computer programmes, they are comparatively easy to develop. As the details of the estimation procedure are somewhat involved, we will not repeat them. There is, however, a complete and lucid survey of these estimation methods available elsewhere (Zellner and Lee (1965)). Instead of dwelling upon the estimation problem, we shall proceed directly to the discussion of an illustrative application.

Application of Probability Models to a Road Construction Problem
The application selected to illustrate the probability model approach concerns common excavation for road construction. The possibility of performing this construction activity is of course highly related to weather conditions. The model presented was developed to provide probabilistic forecasts of the possibility of engaging in the excavation construction activity on the days during the summer season. Results of such investigations are of use in planning and management decisions of highway agencies and construction firms. Together with weather records and forecasts they can be used in day-to-day operations and in longer-run decisions concerning contract delays, acquisitions of capital equipment and the like (Attanasi *et al* (1972)).

With this brief introduction we present estimates of work-day probability equations based upon data compiled from records of a State Highway Department. The estimation method utilised in this case was

Table 12.4

LOGIT ESTIMATES OF LINEAR FUNCTIONS RELATING CLIMATICALLY ORIENTATED VARIABLES TO WORKING DAY PROBABILITIES FOR COMMON EXCAVATION

Equation	Constant	7 day average precip	Aver daily temp	Sq of 7 day aver precip	Sq of aver daily temp	Cross prod aver temp; 7 day precip	4 day average precip	Sq of 4 day aver precip	Log of aver daily temp	3 day average precip and present precip	Sq of 3 day average precip and present precip	0·1 var daily precip	3 day average precip
A	—8·3522												
B	—0·104	—0·1031	—0·0374										
C	3·7984	—0·3285	0·0484	0·0036	0·0005	0·0015							
D	13·5450								3·7840				
E	—5·1275	—0·0242		0·0037					1·9065				
F	—13·3656						—0·5232	0·0095	3·8237	—0·0640	0·0002		
	—7·0290								2·3464			—3·0658	—0·0613

the logit technique. Equations A–F in Table 12.4 include the logit parameter estimates for the coefficients associated with the indicated climatic variables. It is of interest to note that the signs for the coefficients are as would be anticipated. Positively signed variables increase forecasted probabilities that excavation activities are feasible whilst negative values have opposite interpretations. Moreover, the equations seem to fit the data rather as measured by the magnitude of the residual variation. Although these measures of fit are not presented, they were all equivalent to coefficients of determination of 80 per cent or above in standard regression problems.

Since the equations are conceptually the same as those employed in standard regression analysis and the forecasted values, $\hat{y}$'s, fall within the unit interval, they may be treated as predictors of conditional probabilities. That is the conditional expectation of the variable y given observed or hypothesised values for the weather variables is the expected probability that the construction activity can take place. As this illustration hopefully indicates, this method has wide possibilities for application. Given current data resources and management and planning models, the method of estimation should prove useful in the development of more precise measures of the economic consequences of the weather.

References

ANDERSON, J. W. *An Introduction to Multivariate Statistical Analysis* (New York, 1958). John Wiley & Sons, Inc.

ATTANASI, E. D., JOHNSON, S. R., LEDUC, S. and McQUIGG, J. D. 'Forecasting Work Conditions for Road Construction Activities: An Application of Alternative Probability Models', *Monthly Weather Review*, **101**, No 3 (1972).

BENSON, F. J. and JOHNSON, S. R. 'Principal Components and Problems of Measuring Economic Relationships Which Include Climatic Variables', Proceedings, Second National Conference on Weather Modification, American Meteorological Society (1970), 419–26.

GOLDBERGER, A. S. *Econometric Theory* (New York, 1964).

HILLER, F. S. and LIEBERMAN, G. J. *Introduction to Operations Research* (San Francisco, 1967).
JOHNSON, S. R., REIMER, S. C. and ROTHROCK, T. P. 'Principal Components and the Problem of Multicollinearity', *Metroeconomica*, **24,** No 2 (1972).
JOHNSON, S. R. and HAIGH, P. A. 'Agricultural Land Price Differentials and Their Relationship to Potentially Modifiable Aspects of the Climate', *Review of Economics and Statistics*, **52,** No 2 (1970), 173–80.
MAUNDER, W. J. *The Value of the Weather* (1970).
MCQUIGG, J. D., JOHNSON, S. R. and TUDOR, J. R. 'Meteorological Diversity Load Diversity, a Fresh Look at an Old Problem', *J of Applied Meteorology*, **11,** No 4 (1972), 561–6.
ZELLNER, A. and LEE, T. H. 'Joint Estimation of Relationships Involving Discrete Random Variables', *Econometrica*, **33,** No 2 (1965), 382–94.

CHAPTER 13 W. J. MAUNDER

National Econoclimatic Models

Introduction

In many areas, the emergence of more sophisticated and often more weather-sensitive systems has created a need for more rational and effective responses to atmospheric events. This need for better and more useful information about the atmosphere has come about as a result of three important concepts, developed in part through the work of Sewell (1968), Maunder (1970), and McQuigg (1972). They are:

(1) That the atmosphere is an important natural resource which may be perceived, tapped, modified, despoiled, or ignored, and its resource-availability may be forecast.
(2) That information concerning the atmosphere, such as the past weather and climate, the present weather, and the future weather and climate, is also an important resource.
(3) That, given an understanding of physical-biological-sociological interactions with the atmosphere, and given sufficient information about atmospheric events, man can at times use his management ability to improve the economic and social outcome of many weather-sensitive activities.

There is, in fact, an increasing awareness that weather, and information about the weather, can play a very important role in the decision-making processes associated with the management of 'weather-sensitive enterprises'—see, for example, Perry (1971). Included in these enterprises are aspects of national economies concerned with productivity, such as the total retail trade, total milk production, and total electric power consumption, which are the concern of 'high-level' decision-makers in the form of national governments, or the directors of national companies. The question whether meaningful relationship can be obtained between weather indices and economic activities over such space scales may be validly put; however, the continued publication of 'weekly and monthly indicators of economic activity' in business journals such as *Business Week* and *The Economist* is an indication that 'high-level' decision-makers are interested in these data, and since decisions are undoubtedly made on the basis of these data, it appears not unreasonable to attempt to incorporate into the decision-making process *one* of the environmental factors which is associated with these national economic indicators—that of the weather. Assuming, therefore, that a relationship does exist between nation-wide economic activities and nation-wide weather, the immediate problem is to establish its magnitude, and, in order to do so, the problems of establishing such a relationship must be critically examined.

An argument could, of course, be presented that any relationships derived from weather-economic studies cannot be used until 'reasonably accurate' weekly or monthly weather forecasts are available. However, it is assumed that such forecasts will become available and that decisions based on such forecasts will or should be made, for the alternative is that weekly and/or monthly forecasts, as at present issued in the UK and USA and other countries, are just fun and games. Moreover, if meteorologists are serious about predicting the weather, then one justification for more effort in weather prediction is the potential use of this information by 'consumers' of weather forecasts, including decision-makers at national and regional levels.

The Value of Weather to an Area
Irrespective of any long-term climatic trend, by far the most significant weather variations occur from day to day, week to week, and month to month, and it is with these relatively short-term variations that

decision-makers of weather-sensitive enterprises have to contend. The value that can be placed on these climatic resources, in terms of the variations that occur in the short-term 'weather', requires, however, the identification of activities in that area that are affected directly or indirectly by weather changes, and an analysis of the manner in which a given change results in gains or losses to such activities. Specific aspects of this problem have been considered by a number of investigators, including Musgrave (1968) on housing starts, McQuigg, Johnson and Tudor (1972) on electric power consumption, McQuigg and Thompson (1966) on natural gas consumption, Johnson and Haig (1970) on agricultural land prices, Maunder (1968) on agricultural incomes, and Russo (1966) on the construction industry. In most cases, however, these investigations have either referred to a restricted area, or they have been related to a period of at least a month. Such studies are not therefore designed to contribute to the more general question of 'what is the value of weather over a large area and for a short period of time?'.

An important associated aspect of the value of climatic resources is the increasing interest in the benefits obtainable through a better use of weather information, and studies by Maunder (1970, 1971), Sewell *et al* (1968), McQuigg (1970, 1972), Taylor (1970, 1971), and Maunder, Johnson and McQuigg (1971a, b) have shown that proper climatological advice can play an important role in the decision-making involved in a weather-sensitive enterprise. Attempts to find associations between economic and climatic data, however, present many problems including the incompatibility of the two raw data sets, since economic information is related to areas and climatic data to places. Most national meteorological services, for example, publish climatic information for places, or points, this emphasis on point climates being accompanied by an apparent reluctance to produce climatological data which are applicable to areas, and in particular data which are applicable to economically important areas. Because of this, it is usually necessary for the economic-climatologist to transform either the economic or the climatic data, and from the nature of the information, climatic data generally are the easier to adjust.

As a contribution to the problem of developing national econoclimatic models, this paper now discusses (1) the formulation of weighted rainfall, temperature, and water deficit indices on a monthly basis for New Zealand, and their application to New Zealand dairy production and New Zealand electric-power consumption, and (2) the application of weighted weekly precipitation and temperature indices to weekly retail trade sales in the United States.

The New Zealand Example

The problems associated with the development of economically important climatic indices through econoclimatic models is discussed in detail elsewhere (see Maunder, 1972a, 1972b), but an essential part of these models is the calculation of weighted weather indices based on the contribution of various areas (counties) to the total 'economic activities' (such as human population, dairy cow population) in a region. In New Zealand, the basic economic unit for statistical purposes is the county ('geographical' or 'administrative'), and although many county boundaries do not coincide with geographical boundaries, the counties in New Zealand can be used as the basis for a regional breakdown of economic activities. Indeed, in many instances, the county is the only data source which can be used for a regional division of New Zealand. In 1969, there were 110 counties in New Zealand, and the 'economic' data from these counties—and the climatic data applicable to one or more climatological stations in each county—were used as the basis for the formulation of weighted weather indices.

The calculation of the weightings of the climatic data applicable to the climatological stations used in the analysis is based on the contribution of the county to the New Zealand total 'population' or 'area'. For example, if New Zealand is divided into 21 regions, together with a 'region' for the North Island, the South Island, and for New Zealand as a whole, then, within each region, each county can be considered as contributing a proportion of the total economic activity of the region. Economically important weighted weather indices for each region can therefore be calculated by 'allocating' the county economic parameter (such as those given in Table 13.1) to an appropriate climatological station in each county, and then computing a weighted weather index for each region.

Table 13.1

RELATIVE SIGNIFICANCE OF VARIOUS DATA FOR FOUR GEOGRAPHICAL COUNTIES IN NEW ZEALAND*—DATA INDICATES PERCENTAGE OF NEW ZEALAND TOTAL

Economic parameter	Geographical county†				New Zealand total
	Waikato	Hutt	Waimate	Southland	
Human population	1·0	11·1	0·3	3·3	2,677,000
Land area	0·6	0·6	1·4	3·6	103,000 sq miles
Sheep population	0·8	0·4	1·9	8·9	59,940,000
Dairy cow population	5·5	0·3	0·1	0·9	2,232,000
Beef cattle population	1·3	0·4	0·7	2·4	4,549,000
Crop area	0·4	0·1	1·4	11·4	1,339,000 acres

* Based on data applicable to the 1966–9 period.
† As defined by the Department of Statistics.

In the following analysis, three climatological elements (monthly rainfall, monthly mean temperature, days in a month with water deficit) are used to obtain three weighted climatic indices, the weightings used being (1) land area, (2) human population, (3) sheep population, (4) dairy cow population, (5) beef cattle population, and (6) crop area. Accordingly, for each month assessed, six weighted indices were computed for each of rainfall, mean temperature, and days with water deficit.

In the first case actual monthly rainfalls at 109 climatological stations were expressed as a percentage of the 1921–50 normal rainfall for the specific stations. These percentages were then weighted by multiplying them by the county contribution of the New Zealand total population or area. In the second case, the actual number of days per month with a 'water deficit' (based on a Thornthwaite assessment of 'water deficit' and assuming a soil moisture capacity of 76mm) were weighted using the same method. Thirdly, weighted monthly temperature departures from the 1931–60 temperature normals were calculated based on the weightings of 78 climatological stations.

The weighted county climatic indices for each month were next combined into 21 regional climatic indices together with indices for the North Island, the South Island, and New Zealand, using the following equation:

$$\text{Climatic Index } I = \frac{\Sigma C_i E_i}{\Sigma E_i} \qquad \textit{Equation 1}$$

where:

C_i is either the rainfall for station i, expressed as a percentage of 1921–50 normal, the mean temperature departure for station i from the 1931–60 normal, or the actual number of days with water deficit (per month) for station i; E_i is the percentage of the New Zealand economic parameter in county i; and I is a climatic index which ranges from 0 to over 200 in the case of rainfall, $\pm x \cdot x°$ C in the case of temperature departure, and 0–31 days in the case of days per month with water deficit. The indices may therefore be considered indices of 'warmth', 'coolness', 'wetness' or 'dryness'.

Monthly climatic indices for rainfall, mean temperature, and days of water deficit were then computed for each month over the last several years (a selection of the indices for New Zealand for each month from January 1969 to April 1970 for three economically important weightings is shown in Table 13.2). It is believed that such indices are of consider-

Table 13.2
WEIGHTED CLIMATIC INDICES FOR NEW ZEALAND

Year	Month	Climatic index Rainfall* (100 = normal)	Water deficit† (days)	Temperature‡ (° C)
1969	January	103	1·8	−0·1
	February	92	2·3	−0·6
	March	59	7·2	+0·1
	April	97	3·7	−0·6
	May	93	0·1	+0·3
	June	73	0·1	−0·8
	July	56	0·1	−0·7
	August	78	0·1	+0·3
	September	124	0·1	+1·5
	October	67	0·1	−0·9
	November	61	2·3	+0·9
	December	148	0·5	+1·9
1970	January	85	5·8	+1·9
	February	60	18·3	+0·3
	March	146	6·6	+1·4
	April	62	0·4	+0·9

* Weighted by land area.
† Weighted by dairy cow population.
‡ Weighted by human population.

able value in assessing the overall 'wetness' or 'dryness', or 'warmth' or 'coolness' of New Zealand in any month.

Applications of New Zealand Model

The primary purpose in developing the weighted climatic indices is to provide a means by which a vast amount of climatological data can be conveniently expressed as an index which is both representative of the climatic data and also meaningful to studies relating climatic events to economic activities. The application of such indices to economic activities, however, is not as easy, and despite popular opinion 'government economic statisticians' can, only on rare occasions, provide the economic-climatologist with the information he desires. In fact, to quote McQuigg (1972):

> There are not as many well-documented, quantified relationships between weather events and economically important activities as one would think existed. One of the great difficulties that has led to this situation is the lack of homogeneous samples of operational data comparable to the reasonably complete body of meteorological observations that exist in many countries.

The basic handicap to applying weather information to nation-wide economic activities in New Zealand and many other countries is, therefore, the almost complete lack of monthly (and weekly) economic indicators. Two notable exceptions in New Zealand are those associated with the consumption of electric power and the processing of butterfat in dairy factories, and national econoclimatic analyses of these two aspects of the New Zealand economy are now briefly discussed.

The butterfat processed by dairy factories in New Zealand for the months January, February, March and April in the years 1966–70 is shown in Table 13.3 together with a 'water deficit' index, weighted according to the population of dairy cows in milk. The 'expected' production in each of the months was assessed graphically, and it is believed that the 'differences from the expected' reflect a true variation in the butterfat processed by dairy factories. As would be expected, there is some lag between the actual weather experienced and production, but as a first approximation the number of days with a water deficit in the month of production and in the two preceding months can be used as a 'wetness' or 'dryness' index for correlation with the 'difference from the expected' production. Such an analysis (Fig 13.1) indicates a relatively close relationship ($r = -0.82$; significant at 0.1 per cent level) between butterfat production in New Zealand as a whole, and a 'dryness' index for New Zealand as a whole.

It should be noted that more refined analyses on the relationship between climatic factors and butterfat production have been made (see

Table 13.3

BUTTERFAT PROCESSED BY DAIRY FACTORIES IN NEW ZEALAND* AND ASSOCIATED CLIMATIC INDICES FOR NEW ZEALAND WEIGHTED ACCORDING TO THE POPULATION OF DAIRY COWS IN MILK, JANUARY–APRIL: 1966–70

Month	Year	Actual production (m lb)	Difference from expected (m lb)	$(m)†	Water deficit indices (days)			
					Month $(n-2)$	Month $(n-1)$	Month (n)	Month $(n)+(n-1)+(n-2)$
March	1970	30·5	−27·6	−8·3	6	18	7	31
April	1970	20·4	−19·6	−5·9	18	7	0	25
Feb	1970	46·3	−15·9	−4·8	1	6	18	25
March	1968	43·0	−12·4	−3·7	2	14	12	28
April	1968	35·6	− 4·7	−1·4	14	12	1	27
Jan	1970	72·4	− 4·6	−1·4	2	1	6	8
April	1969	36·2	− 4·5	−1·4	2	7	4	13
Feb	1968	56·4	− 2·2	−0·7	0	2	14	16
Jan	1967	70·8	− 0·8	−0·2	0	0	1	1
April	1967	40·0	− 0·8	−0·2	1	2	2	5
March	1969	56·5	− 0·5	−0·2	2	2	7	11
Jan	1966	68·2	− 0·4	−0·1	0	1	1	2
April	1966	37·1	− 0·2	−0·1	2	2	1	5
Feb	1969	60·2	− 0·2	−0·1	0	2	2	4
Jan	1969	75·9	− 0·2	−0·1	0	1	2	3
Feb	1966	54·3	+ 0·4	+0·1	1	1	2	4
Jan	1968	75·0	+ 0·9	+0·3	0	0	2	2
March	1967	54·3	+ 1·3	+0·4	1	1	2	4
March	1966	51·7	+ 1·5	+0·5	1	2	2	5
Feb	1967	57·9	+ 1·6	+0·5	0	1	1	2

* In order of difference from the 'expected' production.
† Based on a net return to farmers of 30 cents per lb.

for example: Curry (1958), McMeekan (1960), Maunder (1968)), but in these studies the productivity and climatic factors were primarily concerned with small areas, whereas the purpose of developing indices as described in this paper is to provide a means of associating production with climatic factors on a national scale.

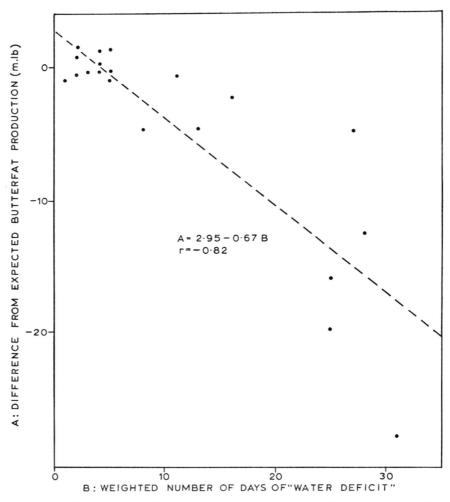

Fig 13.1 Monthly New Zealand butterfat production and associated weighted 'water deficit' indices, January–April, 1966–70

A graphical analysis of the relationship between the monthly temperature departures for New Zealand (weighted according to the distribution of the human population), and the random oscillation of the total monthly electricity generation in New Zealand (as determined by the Department of Statistics, after allowance for seasonality, trend, and 'work' days) is given in Fig 13.2. The analysis covered the 60 April to September months in the 1961–70 decade, and indicated that 25 per cent ($r = -0.50$; significant at the 0.1 per cent level) of the variance in the random oscillation of New Zealand monthly electricity generation was associated with the departure of the weighted monthly New Zealand temperature indices from the 1931–60 normals. Such an analysis provides therefore a valuable insight into the temperature sensitivity of an important sector of the New Zealand economy.

The United States Example

In this example the formulation of weighted indices on a weekly basis for the United States is discussed. The factors of 'large area' (ie, the United States) and 'short time period' (ie, a week) are chosen deliberately for two reasons: first, it is believed that nation-wide economic activities are important to various 'high-level' decision-makers, and second, only weather over a short period (such as a week) has any real practical meaning to the 'low-level' decision-makers in the form of the many millions of Americans who each day go shopping. It could of course be argued that a weather index for a nation as large as the United States has little physical or practical meaning. Nevertheless, it is strongly believed that some measure of nation-wide weather can be computed, and that it can be of use to decision-makers. The alternative is that business, including business on a nation-wide scale, may choose to ignore an important aspect of the environment—the weather—an omission which could lead to incorrect decisions with unfavourable economic and social consequences to a nation.

A prime reason for this particular research into weather-economics therefore, is to enable nation-wide economic activities to be related to

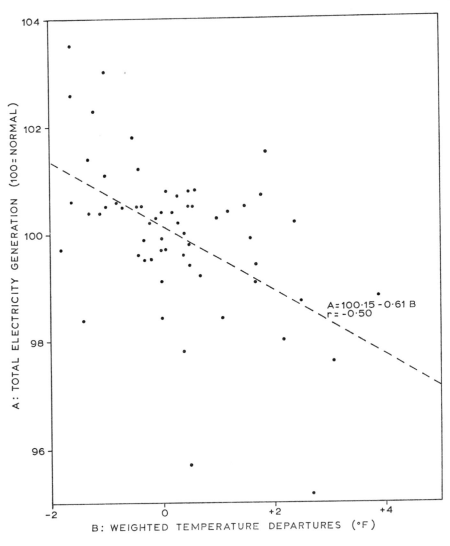

Fig 13.2 Monthly New Zealand electricity generation and associated weighted temperature departures, April–September, 1961–70

nation-wide weather, a specific aspect of the nation-wide weather being the data published weekly by NOAA in the *Weekly Weather and Crop Bulletin*, as well as the five-day nation-wide weather forecasts issued by the National Weather Service.

Various kinds of weekly economic indices are available for the United States; among these are those published by the US Department of Commerce in their *Weekly Retail Sales*, and these data formed the basis of this investigation. It should be emphasised that most of these and other similar sets of weekly data are available only on a nation-wide basis. Consequently, irrespective of the value of such data on a regional scale, the fact that they are not available necessitates the calculation of national weather indices on a weekly basis, if reasonably meaningful econoclimatic studies on a weekly basis are to be made. It should also be noted that the official publication *Weekly Retail Sales* of the US Department of Commerce, which gives estimated weekly retail sales for the United States, is restricted to 13 major *store types*, and unfortunately no weekly information is published relating to the weather-sensitive *commodities* such as ice cream, soft drinks, women's winter and summer clothes, refrigerators, air-conditioners, automobile tyres, beer, iced tea, paint, umbrellas, etc.

Despite the limitations of the economic data, the computation of associated weather data on a comparative basis (ie for the United States as a whole on a week to week basis) was considered justifiable. A problem that arises immediately is that people react to weather rather than to the weather elements of rain, snow, temperature, wind, humidity or sunshine. But, as will be appreciated, the combination of such weather elements into a single weather index is difficult, and although arguments could be advanced for formulating an index following the methods used in agroclimatology (such as de Martonne's or Angstrom's aridity index described by Oury (1965)), the problem is still formidable. Moreover, since the only suitable published weather data on a weekly basis for stations in the United States is that in the *Weekly Weather and Crop Bulletin*, it is justifiable in the first instance to use these data—notably

R

precipitation and temperature—bearing in mind that a combination of the wind, sunshine, humidity, temperature, and precipitation duration would probably be more appropriate.

In order to compute a nation-wide rainfall and temperature index, indices were evaluated from 'weighted' weekly mean temperature and weekly precipitation departures from normal for 147 places, most of these places having a 1965 metropolitan population of 300,000 or more, the weighting being computed for each place based on the 'buying power index' as published in the *Marketing Magazine* (10 June 1969) '1969 Survey and Buying Power', as follows:

$$\text{Buying Power Index} = (5I + 3R + 2P)/10$$

where:
- I = percentage of US Effective Buying Income
- R = percentage of US retail sales
- P = percentage of US population

The resulting weighted temperature and precipitation departures were then combined into indices for the United States using the following expression:

$$\text{United States Climatic Index } I = \frac{\Sigma W_i E_i}{\Sigma E_i} \qquad \text{Equation 2}$$

where:
- W_i = the rainfall or temperature departure from the average for station i ($i = 1 \ldots 147$)
- E_i = the 'buying power index' for station i ($i = 1 \ldots 147$)

An abridged example of the calculation of the weighted temperature and precipitation departures from the normal for the United States for the week ending 12 February 1968 is given in Table 13.4. Similar computations were made for each of the 154 weeks during the 3 year period from April 1966 to March 1969. The association between retail trade sales and the weighted precipitation and temperature indices for the United States (sample data for seven weeks in January and February

Table 13.4

ABRIDGED EXAMPLE OF THE CALCULATION OF TEMPERATURE* AND PRECIPITATION† DEPARTURES FROM THE NORMAL FOR THE UNITED STATES FOR THE WEEK ENDING 12 FEBRUARY 1968, WEIGHTED ACCORDING TO THE 'BUYING POWER INDEX' OF 147 LOCALITIES

No	Station	'Buying power index' (US=100)	Departure from normal temp (°F)	Departure from normal precip (in)	Weighted differences temp (°F)	Weighted differences precip (in)
1	Birmingham, Alabama	0·32	− 9	−1·3	− 2·8	−0·42
19	Washington, DC	1·67	− 2	−0·5	− 3·3	−0·84
22	Miami, Florida	0·63	− 8	−0·2	− 5·0	−0·13
32	Chicago, Illinois	4·22	− 2	−0·3	− 8·4	−1·23
50	New Orleans, Louisiana	0·50	−11	−1·0	− 5·5	−0·50
65	Kansas City, Missouri	0·76	− 3	−0·3	− 2·3	+0·23
72	Las Vegas, Nevada	0·14	+ 3	+0·1	+ 0·4	+0·01
80	New York, New York	9·27	− 4	−0·8	−37·1	−7·42
122	Dallas, Texas	0·79	− 1	−0·6	− 0·8	−0·47
132	Salt Lake City, Utah	0·26	− 4	−0·3	+ 1·0	−0·08
147	Cheyenne, Wyoming	0·03	+ 7	−0·1	+ 0·2	−0·00
	Total (147 stations):	63·12	—	—	−152·8	−35·35

* Total weighted temperature departure = −152·8.
Average departure = −152·8/63·12 = −2·4° F (−1·3° C).
† Total weighted precipitation departure = −35·35.
Average departure = −35·35/63·12 = −0·56in (−14·2mm).

1968 are given in Table 13.5) were then assessed for various 11 week periods using the following equation:

$$x_1 = a_0 + a_1 x_2 + a_2 x_2^2 + a_3 x_3 + a_4 x_4 \qquad \text{Equation 3}$$

where:

x_1 = retail sales for the specific kind of retail business for specific weeks

x_2 = time factor (week 1, 2 ... 11)

x_3 = weighted precipitation departure for the United States for specific weeks

x_4 = weighted temperature departure for the United States for specific weeks

Table 13.5

SELECTED WEEKLY RETAIL TRADE SALES* IN THE UNITED STATES FOR SEVEN WEEKS IN JANUARY AND FEBRUARY 1968 AND ASSOCIATED PRECIPITATION AND TEMPERATURE INDICES WEIGHTED ACCORDING TO THE 'BUYING POWER INDEX' OF 147 LOCALITIES

Period: week ending Sat	Kind of retail business (millions of dollars)					Weighted indices†	
	Total retail trade	Apparel group	Furniture and appliances	Lumber building hardware	Drug stores	Precip (in)	Temp (°F)
Jan 13	5,344	284	250	227	205	+0·10	−8·5
Jan 20	5,562	296	281	255	215	−0·32	+1·9
Jan 27	5,581	273	276	264	208	−0·28	+1·8
Feb 3	5,706	277	295	275	203	+0·28	+6·4
Feb 10	5,720	292	287	286	210	−0·56	−2·4
Feb 17	5,772	264	278	297	227	−0·43	−5·8
Feb 24	5,778	284	302	287	200	−0·48	−6·3

* Weekly sales estimates are based on data from 2,500 firms, covering approximately 48,000 retail stores in the United States. *Source of data: Weekly Retail Sales Report*—US Department of Commerce/Bureau of the Census. *Note:* Data are not adjusted for seasonal or holiday variations.

† Weighted weather indices are for week ending midnight on the Sunday of the week indicated. The indices shown are departures from the normal for the specific week.

The weekly retail trade data as published are *not* adjusted for seasonal or holiday variations, and in view of this the 154 weeks' data were grouped into 11 week overlapping periods. The 11 week periods were centred at the mid-point of the months February to November inclusive in each of the three years. In considering the various 11 week periods certain holiday weeks were excluded, these weeks being replaced by the nearest non-holiday week outside the 11 week period. In addition, because of Christmas, the analyses were not extended into December.

The regression equation is of a form which has been used extensively in agroclimatology—see, for example, Oury (1965), Doll (1967), and Thompson (1962), the equation in this case allowing an assessment of the relationship between variations in retail trade sales from a quadratic time trend (over the specific 11 week periods), and the variations in the nation-wide precipitation and temperature indices. For example, the following is the computed regression equation for the 11 week period 13 January 1969 to 24 March 1969 for retail sales in the apparel group:

$$x_1 = 357\cdot31 - 26\cdot23\ x_2 + 2\cdot58\ x_2^2 - 60\cdot10\ x_3 + 4\cdot64\ x_4$$

Equation 4

The mean value of x_1 in this period was $313 million, the regression giving a standard error of the estimate of $11 million, an R^2 value of 0·94 and an F ratio of 22·2, indicating significance at the 0·1 per cent level. The inter-relationships between first, the weighted precipitation departure (x_3) and the retail sales (x_1), and second, the weighted temperature departure (x_4) and the retail sales (x_1), were then examined using the relevant partial correlations. Analyses were made for all the 11 week periods (except December) in the three years for all 13 aspects of the retail trade.

Because of the small size of the sample, the precipitation/retail trade and temperature/retail trade correlation coefficients for each corresponding 11 week period in the three years were combined, by converting the r values to Fisher's z' values, finding a weighted mean of the z' values, and then converting back to a new r value (see Brooks and Carruthers (1953, 222)).

Weather-Retail Trade Associations

An analysis of the resulting partial correlations shows that in the period February to June, 14 of the 65 precipitation partial correlations were significant at the 10 per cent level, compared with an expected 6·5. This difference, according to the Binomial distribution and assuming that the variates are independent, has a chance occurrence of about 1 in 100. Similarly in the case of the temperature partial correlations, 15 of the 65 March to July correlations were significant at the 10 per cent level, a result which is also unlikely to arise by chance. A more detailed analysis of the partial correlations (sample correlations are given in Table 13.6), however, indicates the probable greater importance of temperature as a factor associated with retail trade sales, and also the relatively close association of late winter to early summer weather conditions with retail sales variations, compared with the absence of significant associations in August and September.

As would be expected with weather and economic data 'averaged' over a large and non-homogeneous area such as the United States, the econoclimatic associations revealed by the partial correlations vary in their significance. Indeed for some retail trade groups, and during some of the 11 week periods, the indicated relationships have very limited practical value. However, it is believed that the partial correlations do give a valuable insight into the overall relationships between weekly weather conditions on a national scale and the published weekly retail sales on a national scale, particularly in regard to the seasonal sensitivity of various kinds of retail sales to precipitation and temperature variations from the normal. The analysis in fact gives emphasis to the generally held view that many aspects of the variations in the retail trade in the United States are associated with the lateness or otherwise of the winter and summer seasons.

Conclusion
An important factor in any national econoclimatic study is that most economically associated information is related to areas, whereas most published climatic data applies to points. It is believed that the method of adjusting weather and climatic data to fit various kinds of nation-wide economic data, as described in this paper, offers a solution to the data problem confronting studies in weather economics.

As has been emphasised, it would be very desirable if weekly regional data, for the economically important activities of various nations, were

Table 13.6

SAMPLE ASSOCIATIONS* BETWEEN WEIGHTED PRECIPITATION AND TEMPERATURE DEPARTURES AND TOTAL RETAIL SALES FOR VARIOUS KINDS OF BUSINESS

Kind of business	March	April	May	June	October
(a) *Precipitation*					
Chain grocery stores	+0·05	+0·10	+0·42†	+0·05	−0·19
Department stores	+0·38†	+0·39†	+0·45‡	+0·34	+0·04
Apparel group	+0·24	+0·06	+0·40†	+0·35	+0·18
Furniture and appliances	−0·18	+0·05	+0·24	+0·23	−0·45‡
Lumber, hardware, farm equipment	+0·58§	+0·07	−0·24	−0·08	−0·38†
Drug and proprietary stores	−0·36	+0·18	+0·25	+0·35	+0·20
(b) *Temperature*					
Eating and drinking places	+0·03	−0·14	+0·13	+0·51‡	+0·08
Department stores	+0·50‡	+0·17	+0·36	+0·36	+0·15
Apparel group	+0·46‡	+0·06	+0·24	+0·25	0·00
Furniture and appliances	−0·01	+0·38†	+0·41†	+0·62§	+0·61§
Automotive group	+0·45‡	+0·56§	+0·24	+0·26	−0·08
Gasoline service station	−0·07	+0·01	−0·13	+0·12	+0·14

* Based on 11 week period centred on the months shown. Data applies to the 1966–9 period. The associations shown are partial correlations (see text for details).
† Significant at the 10 per cent level.
‡ Significant at the 5 per cent level.
§ Significant at the 1 per cent level.

available for use in econoclimatic studies, but since in most, if not all countries this is not the case, it is necessary to devise methods to incorporate nation-wide economic data into econoclimatic studies. It is considered that the results of such studies are significant to decision-making at various national levels, the alternative being the omission of meteorological conditions from decision-making, an omission which can lead to incorrect decisions with unfavourable results for the economy.

The methods of adjusting weather data used in this paper are only some of a large number of possible adjustments. Additional 'weightings' could also be evaluated and these could provide useful nation-wide weather indices for such additional economic activities as forest protection, crop production, road construction, building construction, communications, electric power generation, transportation, and tourism.

If climatic information is to be used effectively in the decision-making process, it is essential that it be in a form that is appropriate to the user. It is believed that the models described here for the United States and New Zealand provide a useful method of translating weather information into more meaningful indices which may be used with profit by the 'high-level' decision-makers of the more weather-sensitive aspects of national economies.

References
BROOKS, C. E. P. and CARRUTHERS, N. *Handbook of Statistical Methods in Meteorology* (HMSO, 1953).
CURRY, L. *Climate and Livestock in New Zealand—A Functional Geography*, PhD Thesis, University of New Zealand (Auckland, 1958).
DOLL, J. P. 'An Analytical Technique for Estimating Weather Indexes from Meteorological Measurements', *Journal of Farm Economics*, **49** (1967), 79–88.
JOHNSON, S. R. and HAIG, P. A. 'Agricultural Land Price Differentials and Their Relationship to Potentially Modifiable Aspects of the Climate', *The Review of Economics and Statistics*, **52** (1970), 173–80.
MCMEEKAN, J. P. 'Grass to Milk', New Zealand Dairy Exporter Newspaper Co (Wellington, 1960).
MCQUIGG, J. D. 'Some Attempt to Estimate the Economic Response of Weather Information', *World Meteorological Organisation Bulletin*, **19** (1970), 72–8.
——. *The Use of Meteorological Information in Economic Development*, prepared for WMO Executive Committee Panel on Meteorology and Economic and Social Development (May, 1972).
MCQUIGG, J. D., JOHNSON, S. R. and TUDOR, J. R. 'Meteorological Diversity-load Diversity, a Fresh Look at an Old Problem', *Journal of Applied Meteorology*, **11** (1972), 561–6.
MCQUIGG, J. D. and THOMPSON, R. G. 'Economic Value of Improved

Methods of Translating Weather Information into Operational Terms', *Monthly Weather Review*, **94** (1966), 83–7.

MAUNDER, W. J. 'Effect of Significant Climatic Factors on Agricultural Production and Incomes: A New Zealand Example', *Monthly Weather Review*, **96** (1968), 39–46.

——. *The Value of the Weather* (1970).

——. 'The Value and Use of Weather Information', *Transactions of the Electric Supply Authority Engineers' Institute of New Zealand*, **41** (1971), 10–20.

——. 'National Econoclimatic Models: Problems and Applications', *NZ Meteorological Service Technical Note* (1972a), 208.

——. 'The Formulation of Weather Indices for Use in Climate—Economic Studies: A New Zealand Example', *New Zealand Geographer*, **28** (1972b),

MAUNDER, W. J., JOHNSON, S. R. and McQUIGG, J. D. 'Study of the Effect of Weather on Road Construction: A Simulation Model', *Monthly Weather Review*, **99** (1971a), 939–45.

——. 'The Effect of Weather on Road Construction: Applications of a Simulation Model', *Monthly Weather Review*, **99** (1971b), 946–53.

MUSGRAVE, J. C. 'Measuring the Influence of Weather on Housing Starts', *Construction Review*, **14**, No 8 (1968), 4–7.

OURY, B. 'Allowing for Weather in Crop Production Model Building', *J of Farm Economics*, **47** (1965), 270–83.

PERRY, A. H. 'Econoclimate—A New Direction for Climatology', *Area* (Institute of British Geographers), **3**, No 3 (1971), 178–9.

RUSSO, J. A. JR. 'The Economic Impact of Weather on the Construction Industry of the United States', *Bulletin of the American Meteorological Society*, **47** (1966), 967–72.

SEWELL, W. R. D. et al. *Human Dimensions of the Atmosphere*, National Science Foundation (Washington, DC, 1968).

SEWELL, W. R. D. 'Emerging Problems in the Management of Atmosphere Resources: The Role of Social Science Research', *Bulletin of the American Meteorological Society*, **49** (1968), 326–36.

TAYLOR, J. A. (ed). *Weather Economics*, University College of Wales, Memorandum 11, 1968 (Oxford, 1970).

——. 'Curbing the Cost of Bad Weather', *New Scientist and Science Journal* (3 June 1971), 560–3.

THOMPSON, L. M. 'Evaluation of Weather Factors in the Production of Wheat', *Journal of Soil and Water Conservation*, **17** (1962), 149–56.

Author Index

NOTE: Authors listed alphabetically at the end of each chapter have been excluded from this index. References to illustrations (Figures) are in italics; references to Tables are prefixed with the letter 'T'. All references per author have been placed in chronological order within the sequences of page references.

Alcock, M. B., 69, 73
Allen, C. W., 29
Anderson, J. W., 226
Attanasi, E. D., 233
Aucliciems, A., 175, 182
Austin, R. B., 73, 95

Baker, O. E., 127
Barnard, C., 87
Beaver, Sir Hugh, 197
Beevers, H., 96
Bennett, M. K., 128, *6.2*
Benson, F. J., 229
Bibby, J. S., 110, 115, 116
Billingsley, D., 179, 180
Birse, E. L., 49, 51, 116
Biswas, A. K., 159, 161, 163, 167
Black, J., 21
Blacksell, M., 180, 181
Blodget, L., 124
Brazell, J. H., 202
Breese, E. L., 88, 105
Brooks, C. E. P., 253
Brouwer, R., 73, 92, 95
Brown, D. M., 128
Bryson, R. S., 203
Buchan, A., 48
Buchanan, R. A., 39
Budkyo, M. I., 203
Bull, T. A., 88
Buller, A. H. R., 126
Burton, I., 41, 175, 182
Bush, R., 118

Canham, A. E., 92
Carruthers, N., 253
Chandler, T. C., 42
Chapman, L. J., 128
Chapman, V., 183, T.9.3
Charles-Edwards, D. A., 96, 105
Codling, P. J., 216
Cooper, J. P., 88, 90, 95, 96, T.4.1, 103
Costes, C., 87
Craddock, W. J., 130
Craxford, S. R., 175
Curry, L., 39, 245

Daday, H., 105
Dawson, G. M., 125
Dawson, R. F. F., 219
Delwiche, C. C., 33
Denmead, O. T., 68, 69
Doll, J. P., 253
Dorfman, R., 162
Dry, F. T., 116
Duckham, A. N., 39
Dunbar, G. S., 121
Duncan, W. G., 68, 71

Eagles, C. F., 105

Fick, G. W., 73
Flinn, J. C., 70
Foster, L., T.9.1, 177

Gage, W. J., 124
Geiger, R., 56, 67
Goldberger, A. S., 233

258

AUTHOR INDEX

Gooriah, B. D., 175, 182
Gouindjee, 87

Haigh, P. A., 227, 239
Hamilton, H. R., 162
Harrison, S. J., 38
Hayter, R., 131
Hector, J., 123
Henderson, J. M., 161
Hershfield, D. M., 151
Hess, M., 51
Hiller, F. S., 232
Hogg, W. H., 38, 109, 112, 119
Howe, G. M., 41
Hugh-Jones, M. E., 38
Hunt, L. A., 100
Hutchinson, G. E., 36

Idso, S. B., 67

Jackson, R. M., 32
Jenkins, I., 202
Johnson, S. R., 39, 225, 226, 227, 229, 239
Jones, G. E., 36, 200
Jones, L. I. R., 128, *6.2*

Kirkby, A. V., 179, 182, 185, 187
Kleinendorst, A., 89, 92, 95
Knoch, K., 117
Koeppe, C. E., 128

Lamb, H. H., *1.2*, T.1.2
Landsberg, H. E., 40
Langbein, W. B., 153, 155
Langridge, J., 88
Lee, T. H., 233
Leonard, W. H., 128
Leopold, A. C., 88
Lieberman, G. J., 232
Longley, R. W., 131
Loucks, D. P., 161, 165
Louis-Byne, M., 131
Lovett, J. V., 73
Lynn, W. R., 161

Maass, A., 159
McCree, K. J., 77
Mackintosh, W. A., 128, 130
Mackney, D., 110, 115, 116
McMeekan, J. P., 245
Macoun, J., 124, 125
McQuigg, J. D., 39, 231, 237, 239
McWilliam, J. R., 88

Manley, G., 48, 49, 50, 51, 53, 55
Martin, J. H., 128
Maunder, W. J., 39, 223, 237, 239, 245
Mera, K., 162
Miller, A. A., 48
Milliman, J. W., 162
Mitchell, K. J., 90
Monteith, J. L., 68, 69, 71, 76, 95
Musgrave, J. C., 239

Nicholson, E. M., 21

Odum, E. P., 36
Oliver, J., 48, 51, 55
Ollerenshaw, C. B., 38
Orcutt, G. H., 159
Osman, A. M., 95
Oury, B., 250, 253

Palliser, J., 123
Patefield, W. M., 73, 95
Paul, A. H., 41
Peacock, J. M., 69, *4.2*, 92, *4.5*, 95
Pearsall, W. H., 48, 53, 59, *2.4*
Penman, H. L., 30, 71, 143
Perry, A. H., 41, 237
Plass, G. N., 29, 32
Proudfoot, B., 38, 126

Rabinowitch, E., 87
Rahman, M. A., 161
Raw, F., 32
Reed, W. G., 128
Reimer, S. C., 225, 226
Reynolds, P. J., 159, 161
Rijtema, P. E., 70, 72
Robson, M. J., 88, 90, 98
Rodda, J. C., 37, 147, 151
Rothrock, T. P., 225, 226
Roy, M. G., 92
Rudeforth, C. C., 61
Rydz, B., 153
Ryle, G. J. A., 98

Sabie, Barbara E., 219
Scorer, R. S., 41, 187, 197, T.10.1
Scott, J. R., 206, 211
Sewell, W. R. D., 39, 237, 239
Sharp, R. G., 149
Sheehy, J. E., 95, T.4.1, 103
Shellard, H. C., 49
Smith, L. P., 38, 51, 53, 115
Stansfield, J. M., 39

AUTHOR INDEX

Stevens, B. H., 161
Stiles, W., *4.2*
Storie, Valerie J., 219
Stringer, E. T., 39
Stupart, R. F., 127
Swan, J. A., 182
Szeicz, G., 69

Tainton, N. M., 96
Tanner, J. C., 206, 207, 211, 220
Taylor, J. A., 38, 39, 40, 49, 50, 56, 239
Thomas, W. L., 39
Thompson, L. M., 253
Thompson, R. G., 239
Tinline, R., 38
Treharne, K. J., 105
Tudor, J. R., 239

Unstead, J. F., 128, *6.2*

Vanderhill, B. G., 123

Wall, G., 179, 180, 186
Wang, Jen-Yu, 71
Warkentin, J., 123, 124, 125
Watson, D. J., 88
Watt, W. R., 95
Weatherly, M-L. P. M., 175, 182
Went, F. W., 88
Williams, G. D. V., 129
Wilson, D., 105
Wit, C. T. de, 68, 71, 73
Woodford, E. K., 87

Zellner, A., 233

Subject Index

NOTE: References to illustrations (Figures) are in italics; references to Tables are prefixed with the letter 'T'. All references per subject have been placed in chronological order within the sequences of page references.

Accidents—road, cost due to weather, 219, 222; data source, 206; due to fog, 208, 215–17, 221; due to rain and wet roads, 208–14, 221; due to slippery roads, 218–19; due to snow and ice, 208, 214; effects of weather on, T.11.1
Advection, 29
Advisory work, in relation to frost incidence, 119
Agriculture, 115, 121, 232; classification of, 109, 110–11
Agricultural, land, area occupied by, 87; land price, 227, 239; effects due to pollution, 195–6, 198, 199; potential, Western Canada, 124–5
Air, space, 41; corridors, 41
Air pollution, 173, 193; absorption from atmosphere, 197; control costs, 194, *10.1*, 198; damage, 194, *10.1*; dispersal, 197; from domestic heating, 197; from electricity generating stations, 197; from general industrial sources, 197; from motor vehicles, 197; thermal, 201
Aitken nuclei, 40
Altitude, effects of, 47–51, 56–9, *2.4*, *2.5*, 67

Amazonia, 32
Atmosphere, as an earth resource, 23–30; as biological and agricultural resource, 30–7; as economic resource, 37–40, 237; as social resource, 40–2; constitution of, 23, T.1.1; functions of, 24; and pollution, *see* Air pollution; upper, 21, *1.1*, 40

Barley, 123
Binomial distributions, 253
Biological, concern for water pollution, 169; effects of air pollution, 200; yield as affected by climate, 65–7, 72
Biosphere, 23, 28, 30, 36, 37, 41
Black areas, 177, 180
Brazil, 38
Business Week, 238
Butter-fat processing in New Zealand, 244, T.13.3

Canada, 38, 121, 163
Canopy structure, 95
Carbon, 24, 167, 189; cycle, 30, *1.4*, 32; -based life, 23, 31–2; monoxide, 196; carbonic acid, 26

SUBJECT INDEX 261

Carbon-dioxide, T.1.1, 23–4, T.1.2, 29, 30; as affects plant growth, 68; cycle, 30, *1.3*; greenhouse effect, 202; industrial, 29, 32
Clean Air Act, 173, 202, 216
Climate, control on plant production, 112, 121; in agricultural classification, 111; and economic activities, 131; and public policy, 121; and pollution, 195, 202; in Scotland, 116
Climatic, change due to altitude, 48–9; change due to man, 18; determinism, 17; fluctuations, 131, T.12.4; groups, definition of, 116; hazards, 123; indices, 242, 250, 251, 253, T.13.2; limits, 111, 121, *6.2*, 126–30; normals, 129, 130, 229
Climatic resources, 65–83, 109; evaluation of, 65, 70–4, 239; management of, 65, 78–83, 87, 155, 237–9; measurement of, 65–70
Climatology, Agro-, 249, 253; Econo-, 17, 39, 131, 237, 249
Cloud seeding experiments, 201
Coalfield communities, 178
Coefficients, correlation, 253; determination, 235
Coffee, leaf-rust disease, 38
Computer, 162
Conifers, 200
Conservation movement, 16, 21
Construction of roads, and climate, 232, 239, 256
Corn belt, of USA, 229
Costs, accident, 219, 222; air pollution, 193, 198; /benefit analysis, 204; economic, 195; hydrological network, T.7.5; smoke control, 176, 188; social, 195; water, 153; waste treatment, 166–7; wheat crop, 122
Crop growth-rates, 70–3, 98, *4.7*, 103–4; problems of selecting suitable measurement of, 90–1
Crop limits, general, 121–2; West Canadian, 122; wheat, 121, 126
Crop production, climatic controls on, 112, 121, 256; potential, 109, 121, 125

Damaging effects of pollution, 194–5
DDT, 25
Decision making, 15, 39, 159, 161, 163, 224, 232, 237–8, 255, 256
Denitrification, 33
Desalination, 153
Domestic heating practices, 175
Dry-matter production, 88, 100, 105

Earliness, of crop production, 114
Economic, activities affected by weather and climate, 223, 256; indicators, 224 (in New Zealand); 249 (in USA); productivity, 18, 21, 37
Econoclimatology, *see* Climatology
Efficiency, light energy conversion, 97–8, T.4.1, 95
Effluent discharge, 167–8
Electric power, as economic indicator, 239, 244, 247, *13.2*, 256; hydro, 135–6, 165
Environmental, crisis, 21; education, 42; management, 17, 23, 42; pollution, *see* Pollution; problems, 183
Equations, climatic index, 242, 250–3; climatic variable, 225; distribution of light within canopy, 96; diurnal variation of light intensity, 96; growth response to temperature, 92; Penman, 71–2; probability, 233; rainfall, general frequency, 15; rate of gross photosynthesis of individual leaves, 96; Rijtema, 72–3
Ethylene, 201
Europe, 32
Evaporation, 28, 137, 143, 147; *see* Plant water potential
Exosphere, *1.1*
Explicit reach, 165
Exposure, 114, 118

Fertilisers, 201
Flood control, 135, 136, 145, 149
Fluorine, 200
Fog, 41; and accidents, 215–17, 221; and pollution, 201–2
Forest, clearance, as affects frost incidence, 125; productivity, as affected by climate, 59, 256
Forestry, 111, 163
Frost, 114, 118–19, 123; free season, 123, 128, 131
Fuel, 175–6, 199

Gas, consumption, 239; as a pollutant, 198; *see also* Carbon, Nitrogen, Oxygen, Sulphur
Genetic variation, 103
Geography, academic, divisiveness in, 15
Geology, effect on climate, 117
Geomorphology, effect on climate, 117
Gliding, 41
Grass production, 65; model for prediction of, 73–4, 78, 95

SUBJECT INDEX

Growing season, 48, 50–1, 58, 73, 125, 128, 229
Growth, gradient, 49–51, 54, 58, 77–8; threshold, 49
Guano, 36

Health hazard, of air pollution, 185, 195, 204
High stacks, 198
Hydro-electric power, 135–6, 165
Hydrogen, 24
Hydrological, appraisal, 135; cycle, 23, 38, 137; forecasting, 153; structures, 136

Indices, buying power, 250, T.13.4, T.13.5; climatic, 242, 250–1, 253, 256
Industrial, pollution, 114, 163, 166, 174; revolution, 40
Ionosphere, *1.1*
Irrigation, 38, 136, 143
Isohyets, UK residual rainfall, *7.3, 7.6*

Katabatic air flow, 119

Land, agricultural, 109, 110; classification, 109, 116; horticultural, 109; prices, 227, 229, T.12.2; scale, 110; systems and grades, 110; -use capability, 109, 110, 113, 121, 123
Lapse-rates, 50–6; construction of, 53–5; definition of, 50; spatial variation of, *2.1*, 28, 52; standard, 50, 55; temporal variation of, *2.2*, 53; validity of, 118
Lead pollution, 189
Leaf, area index, 74, *3.1*, 76, 95, 98, 101; damage, 200–1; expansion, 81, 88, 90–2, *4.3, 4.4*, 100; growth, 88, 90, 92
Least squares method, 233
Light flux density, as affects plant growth, 67, 76
Limestones, 23, 141
Limits, agricultural, 122–3, 126
Liver fluke, 38

Macroclimate, 112, 113, 115, 130
Man, interference by in the biosphere, 32–3, 36, 40–1, 202; and management ability, 237
Mathematical modelling, 159–61
Measurements of, climatic resources, 66; crop growth, 90; damage effects, 197, 203; leaf extension, 90–1; pollution effects, 223; temperature, 69, 89, *4.2*, 90
Mesoclimate, 112, 114, 117–19, 130

Mesosphere, *1.1*
Meteorological, office (MO), 37, 143, 202; device, information and forecasts, 39, 128, 206; stations, 130, 140, 210, 239, 242
Microclimate, 55, 56, 67, 122; measurement of, 67; modification of, 78
Milk production, 238
Models, building of, 15, 224–5, 233; application to road construction, 233; for grass growth, 95; linear probability, 232; microclimate, 79; pollution damage and control costs, 193–4; predicting crop yields, 71–8; model for water resource system planning, 159–61; descriptive, 161; dynamic programming, 159; geometric programming, 159; linear investment, 161; linear programming, 159, 161, 165, 170; predictive, 162; programming, 161; screening, 165
Motorway, accidents, 41; and weather conditions, 216, T.11.6, 220

Natural gas, heating systems, 199; *see* Fuel
Nature Conservancy, 143
New Zealand, 240
Nitrogen, T.1.1, 24, 32; cycle, 33, *1.5*; fertilisers, 33; -fixing, 33; industrial, 33; oxides, as pollutants, 167, 189, 196, 200
Non-linear estimation procedures, normit analysis, 233; logit analysis, 233, T.12.4
North America, 32
Nutrient cycling, 23–4

Oats, 123
Objective functions, 161, 162, 166
Operations research, 159
Optimisation, 166, 168, 171, 199, 204
Oxygen, cycle, 30, *1.3*; dissolved, 166, 169; reservoir, 23, T.1.1, 24, T.1.2, 30
Ozone, 26, T.1.2, 201; concentration, 29; layer, *1.1*, 26

Peat bog, drainage effects, 202
Penman equation, 71–2, 143
Perception, 15, 40–1, 173, 177, 181
Perennial rye-grass, 74, *3.1*, 77–82, 88
Pesticides, 183
Phosphorus, cycle, 34, *1.7*, 36; in fish catches, 36; inorganic, 36; organic, 36
Photosynthesis, 23, 30, 32, 66–7, 71, 76, 79, 87; in the canopy, 96; individual leaf, 96, 103; variation in grasses, 96
Photosynthetic surface, 88

SUBJECT INDEX 263

Physical factors, as affect land capability, 110, 119; see Climate
Planning, 17, 39, 160, 231, 233
Plant-breeding objectives, 105, 126
Plant diseases, black spot, 200; black stem rust, 126; potato blight, 38
Plant water potential, 70, 72, 79, 115–16
Plynlimon catchment, hydrograph characteristics, 7.7, 151–3
PMP (probable maximum precipitation), 149
Pollutants, 24, 28, 40, 168, 189, 193–8, 201
Pollution, 41, 42, 135–6, 173; of cities, 40; of sea, 40, 183; of inland water, 139, 163; industrial, 163, 166; air, see Air pollution
Pre-Cambrian atmosphere, 23
Precipitation, see Climate and Rainfall
Principal components analysis, 224–7
Probability density function, 199
Production, associated with climatic factors, 245
Profitability, 121
Pulp and paper industry, 163

Quaternary studies, 15–16

Radiation, solar, 68, 73, 87; diurnal variation, 4.6, 96; exchange, 24, 1.2, T.1.2, 26, 28, 29; extinction coefficient, 103; measurement, 42; from pollution cloud, 197; utilisation, 105, 112
Radio-active waste, 21, 25
Radio-sonde balloons, 21
Rainfall, and accidents, 208–14, and pollution, 201; inducement, 153; residual, 7.3; total winter, 7.5; variability, 49, 225, 255
Rain gauges, 137, 140, 7.1, 145; instrumental errors, 37
Recreation, 111, 165, 183
Regression analysis, 224, 253
Reservoirs, 135–6, 153, 170
Resources, see Atmosphere, Water
Respiration, 23; as factor in yield prediction model, 77
Response, to air pollution, 177
Retail trade and weather, 254
Rice, short-stem (I.R.8), 38
Rijtema equation, 72–3
River, flow records, 143, 168; routing, 169; simulation, 159; systems, 163
Rough grazing, 111

Safety levels, 21

Salt, deposition of sea salt, 114; spray, 36
Sampling, 163, 173, 181, 232
Satellites, 21
Sciences, behavioural, 15; environmental, 16; management, 159
Scotland, climatic conditions, 117
Seasonal, earliness, 114; lateness, 254; variations, 252
Sewage disposal, 135
Shelter, belts, 38; effect on pasture growth, 79, 3.3, 3.5, 81–3; effect on land-use, 119
Shoot apex, 95, 105
Simulation, computer, 159; model, 165, 168–70
Smog, 41, 182, 202, 216
Smoke concentrations, 9.1, 9.2; control, 173–81, 9.3; effects, 173–81, 200
Smokeless, fuels, 41, 176, 181; zones, 41
Snow and ice, and accidents, 214
Social welfare function, 161
Soil, capability, 229; effects on crop production, 122; effects on mesoclimate, 118; effects on microclimate, 56, 114; heating, effects on growth, 92; productivity, 227; temperature variations, 56–8
Solarimetres, see Radiation
Standard error, 253
Stevenson screen, 50, 55, 74, 88; differences between screen and soil data, 2.3, 89
Stratosphere, 21, 1.1, 26, 29
Sulphur cycle, 34, 1.6, 36; fuel oils, 199; sulphur dioxide, 36, 174, 189, 196, 200
Synthetic flow generation, 168
Systems theory, 15, 159

Temperatures, air, 88–9, 4.2, 4.3, 112, 119, 225, 255; as affects plant growth, 127; effect of crop canopy on, 89, 103; inversions, 188; plant, 69; problems of measurement, 88; profiles, 88–90, 4.2, 4.5, 95; soil, 55–6, 88, 4.2, 112; spatial variation of, 2.1; temporal variation of, 51, 2.2, 79
The Economist, 238
Thermosphere, 1.1
Topography, effects of, 55, 67, 110, 112, 117, 131
Tourism, 256
Toxicity, levels, 25; materials, 42
Traffic, data source, 206; and weather, 205, 208–15, T.11.3, 216
Transfer coefficient, 165

Transpiration, 143; *see* Plant water potential
Transport, *see* Traffic; Transport and Road Research Laboratory, 205
Troposphere, *1.1*, *2.1*

United States of America, 201, 247
Urban redevelopment, 180

Vapour trails, 41
Variable control, 161; policy, 161
Volcanic dust, 29

Waste treatment, 166
Water, deficit indices, 240, 242, 244; demand, 149; ground water, 141, 145, 147, *7.2*; resources of earth, 23; resources of UK, 135; resources planning, 159–70; social quality of, 166–8; supply, 147, 149; vapour, 26, T.1.2, 28, 29

Weather, and air pollution, 201; and architectural adjustments, 40; assessing effects of, 206, 238; costs, 39, 205; economic consequences of, 205, 223, 238; forecasts, 153, 233, 238, 249; hazards, 37, 39; and health, 41; manipulation, 41; and road accidents, 205; and retail trade, 254; sensitive commodities, 37, 249; variations, 338; weighted indices for, 240
Weighting factors, for climatic indices, 197–8, 240–1, 256
Western society, 18, 21
Wheat cultivation, 123, 126; economic viability of, 130; limits of, 121, 126; vernalisation of seed, 38
Wildlife, destruction of, 165, 183
Wind, as a vector in spread of diseases, 38; stress, 40
World, production, 87; agricultural survey, 111